GEOMETRY OF COMPLEX NUMBERS

GEOMETRY OF COMPLEX NUMBERS

Circle Geometry, Moebius Transformation, Non-Euclidean Geometry

HANS SCHWERDTFEGER
Professor of Mathematics
McGill University, Montreal

DOVER PUBLICATIONS, INC.
NEW YORK

Published in Canada by General Publishing Company, Ltd., 30 Lesmill Road, Don Mills, Toronto, Ontario.
Published in the United Kingdom by Constable and Company, Ltd., 10 Orange Street, London WC2H 7EG.

This Dover edition, first published in 1979, is an unabridged and extensively corrected republication of the work originally published in 1962 by the University of Toronto Press. The author has added a new Preface, four Appendices and a Supplementary Bibliography to the Dover edition.

International Standard Book Number: 0-486-63830-8
Library of Congress Catalog Card Number: 79-52529

Manufactured in the United States of America
Dover Publications, Inc.
180 Varick Street
New York, N.Y. 10014

IN MEMORIAM

GUSTAV HERGLOTZ

PREFACE TO THE DOVER EDITION

AFTER THE first appearance of this work in 1962, a number of interesting books and monographs were published (Artzy [1], Benz [2], Davis [1], Magnus [1], Pedoe [2], Yaglom [1], [2])[1], for which my book can serve as collateral or preliminary reading. No major changes have been made; minor errors and misprints have been corrected, some of which have been brought to my attention by readers and reviewers. I am indebted to R. Ping of the University of Kentucky, who pointed out to me a mistake in the formulation of example 4(a) on p. 10. Three appendices have been added in which some recently discovered facts that have bearing on parts of the text are discussed. A fourth appendix draws the reader's attention to the book by Yaglom.

I am grateful to Dover Publications for adding my work to their collection of reprints.

February, 1979 H.S.

[1] These numbers refer to the Supplementary Bibliography (p. 197).

PREFACE TO THE FIRST EDITION

THIS BOOK develops certain sections of a lecture course on the theory of functions of a complex variable which the author has delivered several times during the last twenty years in several universities in Australia and in Canada. Some of its content does not belong to the standard material dealt with in such a course, but was thought to be useful for a deeper understanding of the geometrical theory of analytic functions as well as of the connections between different branches of geometry.

The first two chapters should be understood by a student who possesses a working knowledge of the algebra of the complex numbers and of the elements of analytical geometry and linear algebra, and in addition a willingness to acquire, in the course of reading, the basic ideas of the (non-infinitesimal) theory of groups of transformations, if necessary by occasionally consulting an introductory text of modern algebra.

The third chapter presents the fundamentals of non-euclidean geometry to a student who is primarily interested in algebra and analysis. It is naturally based upon the first two chapters, but requires more specialized interests in the reader.

Each section is followed by a set of examples. Rather than tax the reader's ability, they are designed to convey further information that could not conveniently be incorporated into the main text. In order to invite the reader's collaboration in these examples, simple proofs are left to him or short indications are supplied which should enable him to complete a piece of work in all its details.

The centre of interest of the book lies in the intersection of geometry, analysis, and algebra, in general on a moderately advanced level, with occasional views on recent investigations. Much emphasis, however, has been given to the careful exposition of details and to the development of an adequate algebraic technique.

The Bibliography is not intended to be complete. There are two kinds of references: (1) Books and papers of which use has been made in the composition of the text. These are listed under the name of the author and a number in square brackets. Main references in this category are the books by Caratheodory [3] and by E. Cartan [2][2]. (2) Books and articles (not quoted in the text) containing material related to the subject matter, which might be of interest to the reader.

A first draft of the manuscript was critically revised by my friend Peter Scherk. He supplied me with an extensive list of suggestions and corrections which led to substantial improvements in my exposition. It is with pleasure

[2] Only after the manuscript of the present book was completed in all essential details did I become acquainted with the interesting work by R. Deaux ([1] or [6]). Methods and most of the subject matter in both books are different. Both intend to illustrate the geometrical content of the concept of the complex number system.

that I record here my gratitude to him for all the time and energy he has spent on my work. I am also grateful to Dr. A. Sherk of the University of Toronto who read the final manuscript and to my colleague, Professor D. Sussman, who read the galley proofs; both made valuable mathematical and linguistic suggestions.

Finally I wish to acknowledge the support and encouragement which I received as Fellow of the Summer Research Institute of the Canadian Mathematical Congress at Kingston in 1958, and more generally, as a new member of the Canadian mathematical community. My thanks are due in particular to the editors of the 'Mathematical Expositions' series and to the University of Toronto Press who kindly accepted my book in their series.

H. S.

April, 1960

CONTENTS

GEOMETRY OF COMPLEX NUMBERS

INTRODUCTION

Note on Terminology and Notations

In most branches of mathematics, including the subject to be dealt with here, a general agreement concerning terminology and notation has not been reached and indeed may not be desirable. The author has tried, however, to introduce such names and terms as have been adopted in the majority of writings on the geometry of complex numbers; he has avoided terms suggesting a special interpretation which he did not wish to emphasize. Thus we speak consistently of 'circles,' avoiding the term 'chain.'

The present exposition starts on the basis of analytical geometry using cartesian coordinates x, y of the points in the plane. These coordinates are any real numbers. For our purposes it is expedient to represent the point (x, y) by the complex number $z = x + iy$ where $i = \sqrt{(-1)}$. We call $x = \text{Re } z$ the real part, $y = \text{Im } z$ the imaginary part of z; also $\bar{z} = x - iy$ is the conjugate, and $|z| = \sqrt{(x^2+y^2)} = r \geqslant 0$ is the modulus or absolute value of z. We notice that $z\bar{z} = r^2$ is zero if and only if $z = 0$. The elementary algebra of the complex numbers is assumed to be known.[1]

Using polar coordinates r, θ we may represent the complex number z in the 'trigonometrical form'

$$z = r(\cos\theta + i\sin\theta).$$

We shall call $\cos\theta + i\sin\theta$ the 'angular factor' of z. It is a complex number of modulus one, commonly denoted by $e^{i\theta}$. This exponential notation is justified by the identity

$$e^{i\theta_1}e^{i\theta_2} = e^{i(\theta_1+\theta_2)},$$

that is, the functional equation of the exponential function, which is a formal consequence of the addition theorems of elementary trigonometry.[2] Knowledge of the theory of functions of a complex variable will not be assumed; however, we shall not explain the concepts of continuity and differentiation, which will be used occasionally.

Instead of the angular factor $e^{i\theta}$ of a complex number z we shall sometimes have to introduce the angle θ itself as a function of z; it will be denoted by

$$\theta = \text{arc } z = \text{arc } e^{i(\theta+2k\pi)} \qquad (k \text{ integral}).$$

It is defined for all $z \neq 0$, up to an additive integral multiple of 2π. This multiple may be chosen conveniently from the provenance of the number z. But the range of the values of arc z is often restricted to a certain interval of length 2π,

[1] Cf. R. Courant and H. Robbins [1], chapter II, § 5; or G. H. Hardy [1], chapter III.

[2] Cf. Courant–Robbins [1], pp. 477–9; or Hardy [1], §§ 232–3.

for example, $-\pi < \theta \leqslant \pi$ (the principal value of arc z) or $0 \leqslant \theta < 2\pi$; the function arc z is then discontinuous along the negative or the positive real axis (x-axis) respectively. Its jump value along this axis is $\pm 2\pi$ depending upon the direction in which z crosses the axis.

The basic facts of Linear Algebra are required to an extent not exceeding the amount necessary for an appreciative understanding of the elements of analytic geometry,[3] including homogeneous coordinates (L.A. Appendix, no. 1). We shall make extensive use of the formal laws of matrix algebra (L.A. §§ 12–13); but in the initial stages we shall use only two-rowed matrices whose elements are complex numbers. Such matrices will be denoted by German capitals, for instance

$$\mathfrak{H} = \begin{pmatrix} a & b \\ c & d \end{pmatrix}, \qquad \mathfrak{C} = \begin{pmatrix} A & B \\ C & D \end{pmatrix}.$$

It will be agreed, however, that the symbol $\mathfrak{H}$ stands at the same time for the matrix, as well as for the Moebius transformation

$$Z = \frac{az+b}{cz+d} = \mathfrak{H}(z) \tag{1}$$

which we shall call 'the Moebius transformation $\mathfrak{H}$.' Whereas the Moebius transformation is uniquely defined by the matrix $\mathfrak{H}$, the converse is not true. Given the transformation (1), its matrix $\mathfrak{H}$ is defined apart from a complex number factor $q \neq 0$. Thus the matrices $\mathfrak{H}$ and $q\mathfrak{H}$ representing one and the same Moebius transformation, could have been identified; we have preferred, however, to use only strict matrix equations although in many cases the factor q may remain undetermined.

By $\mathfrak{C}'$ we denote the transpose of $\mathfrak{C}$, viz. $\mathfrak{C}' = \begin{pmatrix} A & C \\ B & D \end{pmatrix}$, and by $\bar{\mathfrak{C}} = \begin{pmatrix} \bar{A} & \bar{B} \\ \bar{C} & \bar{D} \end{pmatrix}$ the conjugate of $\mathfrak{C}$. If $\bar{\mathfrak{C}} = \mathfrak{C}'$, then $\mathfrak{C}$ is said to be a hermitian matrix (L.A. § 20). In § 1 it will be shown that every hermitian matrix $\mathfrak{C}$ defines a circle (in a certain sense) and that $\mathfrak{C}$ and $\lambda\mathfrak{C}$ (λ real $\neq 0$) define the same circle. Thus we shall speak of 'the circle $\mathfrak{C}$.'

The *determinant* of a square matrix $\mathfrak{H}$ will be denoted by $|\mathfrak{H}|$, the modulus of $|\mathfrak{H}|$, however, by $|\det \mathfrak{H}|$.

Transformations play an important role in our exposition. A transformation $Z = f(z)$ is given by a complex-valued function $f(z)$ which associates with every point z in the plane of the complex numbers (or in a certain part of this plane) a complex number Z that appears as a point in the same plane, or in another plane with a congruent coordinate system. We also say that the transformation carries the point z into the point Z (cf. L.A. § 10).

[3] Cf. the early chapters of any text on Linear Algebra, in particular the author's [2], in the following quoted as 'L.A.'

In many cases a set of transformations relevant for a geometrical discussion forms a *group* (cf. L.A. § 22) where the group multiplication is the composition of two transformations of the given set in the sense of forming the function of a function. The unit element in such a group is the identity transformation $Z = z$ which relates every point with itself. Apart from a certain familiarity with the general idea of a group which may be gained by dealing with the special groups of transformations occurring in this book, no elaborate knowledge of group theory is required.

The systems of all rational, all real, or all complex numbers are *fields* in the sense of algebra. This means that within each of these systems the operations of elementary arithmetic—addition, subtraction, multiplication, and division (except by zero)—can be performed so that the sum, difference, product, and quotient of any two numbers of one of these systems are again numbers of the same system. There are many other number fields beside those mentioned above; all of them are contained in the field of all complex numbers and contain in their turn the field of the rational numbers as a subfield (cf. L.A. § 14). In one instance only the reader should realize that fields exist, the elements of which are not numbers, for example, fields consisting of a finite number of elements (L.A. ex. 16.1). In such a field the elementary operations are defined so that the formal laws of arithmetic, namely the existence and uniqueness of the zero and unit elements, as well as negatives and inverses, and the associative, commutative, and distributive laws, are satisfied.

Two fields $\mathscr{F}_1$, $\mathscr{F}_2$ are said to be *isomorphic* (L.A. § 14) if their elements $a_1, b_1, \ldots,$ and $a_2, b_2, \ldots,$ can be put in one–one correspondence, $a_1 \leftrightarrow a_2$, $b_1 \leftrightarrow b_2, \ldots,$ so that $a_1 + b_1 \leftrightarrow a_2 + b_2$, $a_1 b_1 \leftrightarrow a_2 b_2$. To the zero and unit element in $\mathscr{F}_1$ correspond the zero and unit element in $\mathscr{F}_2$ respectively.

An isomorphism of a field onto itself (that is, the case that $\mathscr{F}_1$ and $\mathscr{F}_2$ represent the same field) is called an *automorphism*. All automorphisms of a field form a group. For the field of all rational and the field of all real numbers this group consists of the identity only: $a \leftrightarrow a$ (L.A. ex. 14.8). In the field of all complex numbers the correspondence $z \leftrightarrow \bar{z}$ is readily seen to be an automorphism different from the identity.

Primitive concepts and facts of topology will be used. For explanations and proofs we refer to the literature, mainly books containing introductory chapters.[4] For some examples in chapter III, line and area integrals are assumed to be known. References will be given.

Finally we may point out a number of embarrassing features in the terminology; the difficulties are mainly linguistic in so far as different matters are described conventionally by the same word.

There are different meanings attached to each of the words 'real' and

[4] Caratheodory [3], vol. I, Part II. Courant and Robbins [1], chapter V. Hilbert and Cohn-Vossen [1], chapter VI.

'imaginary' in geometry. The real numbers are the complex numbers associated with the points of (or lying on) the real axis of the complex plane, also called the Gauss plane (or the Argand plane). The pure imaginary numbers are in the same way associated with the real points of the imaginary axis. (Point and associated complex number will always be denoted by the same symbol.) All points in the (complex number, or z-) plane, each characterized by a complex number, are 'real points.' Apart from figures consisting of real points we shall have to consider (only as purely algebraic creations) certain figures, for example, 'imaginary circles' (also called 'ideal circles,' cf. Deaux [1], p. 49) whose (non-real) 'points' have non-real cartesian coordinates; these imaginary circles admit of a real geometrical interpretation only by means of the associated inversion (cf. § 2). Single 'imaginary points' will not occur. Also observe that real circles have imaginary points. A real circle may have a non-real (imaginary) intersection with the real axis; thus there are even non-real points on the real axis!

A word may be said about the use of the adjectives 'hyperbolic,' 'parabolic,' and 'elliptic.' As a matter of fact, there is little justification for these except the reference to the various kinds of non-euclidean geometry where their use is by now historic. So we have hyperbolic, parabolic, and elliptic pencils and bundles of circles as well as motions. But there are also hyperbolic, parabolic, and elliptic Moebius transformations, inversions, and antihomographies, to mention only such objects as will occur in the present book. In these cases the relation to the classical terminology in non-euclidean geometry is more or less remote. General agreement on these definitions seems to exist, however; thus we have accepted them apart from one minor alteration to be explained in § 8, **c**, footnote 4.

I

ANALYTIC GEOMETRY OF CIRCLES

§ 1. Representation of circles by hermitian matrices

a. *One circle.* All points $z = x+iy$ of the complex plane which form the circumference of the circle of radius ρ about the centre $\gamma = \alpha+i\beta$ are characterized by the equation

$$(x-\alpha)^2+(y-\beta)^2 = \rho^2.$$

Writing the left-hand side as $|z-\gamma|^2 = (z-\gamma)(\bar{z}-\bar{\gamma})$ we obtain the equation of the circle in the form

$$z\bar{z}-\bar{\gamma}z-\gamma\bar{z}+\gamma\bar{\gamma}-\rho^2 = 0. \tag{1.1}$$

For our purposes it is advisable to start with the more general equation

$$\mathfrak{C}(z, \bar{z}) = Az\bar{z}+Bz+C\bar{z}+D = 0 \tag{1.2}$$

where A and D are real, and B and C are conjugate complex numbers. The matrix

$$\mathfrak{C} = \begin{pmatrix} A & B \\ C & D \end{pmatrix} \tag{1.21}$$

is therefore a hermitian matrix.

Evidently the equation (1.1) is a particular case of (1.2). Both equations will represent the same circle if $A \neq 0$ and

$$B =- A\bar{\gamma}, \qquad C = -A\gamma = \bar{B}, \qquad D = A(\gamma\bar{\gamma}-\rho^2). \tag{1.3}$$

Every hermitian matrix $\mathfrak{C}$ is associated with an equation (1.2). We shall say that it defines, or is representative of, a 'circle' except if $A = B = C = 0$. Accordingly the letter $\mathfrak{C}$ is used to denote both the circle and the corresponding hermitian matrix. Two hermitian matrices $\mathfrak{C}$ and $\mathfrak{C}_1$ represent the same circle if and only if $\mathfrak{C}_1 = \lambda\mathfrak{C}$ where λ is a real number different from zero.

In order to distinguish between the different types of 'circles' included in this definition we introduce the determinant

$$\Delta = |\mathfrak{C}| = AD-BC = AD-|B|^2, \tag{1.4}$$

evidently a real number, which is called the *discriminant* of the circle $\mathfrak{C}$. For a real circle as represented by the equation (1.1), $\Delta = -\rho^2$. For the circle given by the equation (1.2), according to (1.3) the discriminant is found to be equal to

$$\Delta = -A^2\rho^2. \tag{1.41}$$

Now it is readily seen that the circle $\mathfrak{C}$ given by the equation (1.2) is an ordinary *real circle* if and only if $A \neq 0$ and $\Delta < 0$. Its centre γ and radius ρ can then be found from (1.3), (1.41), and we shall also denote it by (γ, ρ). It will have degenerated into a *straight line* if $A = 0$; in fact, in this case (1.2) turns out to be a linear equation in x and y with real coefficients. For any $A \neq 0$ and $\Delta = 0$ the circle appears as a *point circle*: $\rho = 0$.

Finally we have the case $\Delta > 0$. According to (1.41) this requires that $\rho^2 < 0$. The circle $\mathfrak{C}$ now is an '*imaginary circle*' with a pure imaginary radius ρ and real centre γ. This circle may still be represented by the symbol (γ, ρ) and as in the case of a real circle γ and ρ may be found from (1.3) and (1.41). An imaginary circle has no real point, which means: no point represented by a complex number $z = x+iy$ where x, y are real numbers. As an example we take the 'imaginary unit circle'

$$z\bar{z}+1 = 0, \qquad \mathfrak{C} = \begin{pmatrix} 1 & 0 \\ 0 & 1 \end{pmatrix}. \tag{1.42}$$

Obviously this equation cannot be solved by an ordinary complex number z. It may be written, however, in the form $x^2+y^2 = -1$ and this equation can be solved by 'points' (x, y) whose coordinates x, y are not both real numbers,[1] for example, $x = 0$, $y = i$.

Classification of Circles

$A \neq 0$:	$\Delta < 0$	real circle	$\rho^2 > 0$
	$\Delta = 0$	point circle	$\rho^2 = 0$
	$\Delta > 0$	imaginary circle	$\rho^2 < 0$.
$A = 0$:	Then $\Delta = -\lvert B\rvert^2 \leqslant 0$ always		
	$\Delta < 0$	straight line	
	$\Delta = 0$	no circle: $B = C = 0$ (cf. § 3, ex. 7).	

By (1.41)

$$A = \pm\frac{1}{\rho}\sqrt{(-\Delta)}. \tag{1.43}$$

The real circle $\mathfrak{C}$ is said to have a positive sense of direction or positive orientation if $A > 0$. In this case $-\mathfrak{C}$ will be the same circle with negative orientation. The orientation of a directed circle may be indicated by an arrow on the circumference. Positive orientation corresponds to an anti-clockwise motion on the circumference leaving the interior of the circle at the left side.

The notion of direction or orientation can be extended formally to imaginary circles, the sense of direction being determined by the sign of A.

[1] We are not interested here in a geometrical representation or interpretation of the non-real points of an imaginary (or of a real) circle. Regarding this subject we refer to the book by J. L. Coolidge [2], chapter III.

b. *Two circles.* Let $\mathfrak{C}_1$ and $\mathfrak{C}_2$ represent two different circles; this means that the two hermitian matrices $\mathfrak{C}_1$, $\mathfrak{C}_2$ are not proportional (not linearly dependent) so that there is no real number λ such that $\mathfrak{C}_2 = \lambda\mathfrak{C}_1$. We now study the one-parameter family of circles

$$\mathfrak{C} = \lambda_1\mathfrak{C}_1+\lambda_2\mathfrak{C}_2, \qquad \lambda_1, \lambda_2 \text{ real, not both zero.}$$

It is called a *pencil of circles*. The discriminant of this pencil is, by definition, the determinant

$$(1.5) \qquad |\mathfrak{C}| = \begin{vmatrix} \lambda_1 A_1+\lambda_2 A_2 & \lambda_1 B_1+\lambda_2 B_2 \\ \lambda_1 C_1+\lambda_2 C_2 & \lambda_1 D_1+\lambda_2 D_2 \end{vmatrix} = \Delta_1\lambda_1^2+2\Delta_{12}\lambda_1\lambda_2+\Delta_2\lambda_2^2,$$

that is, a quadratic form in the real variables λ_1, λ_2 with the real coefficients

$$(1.51) \qquad \Delta_1 = |\mathfrak{C}_1|, \qquad \Delta_2 = |\mathfrak{C}_2|, \qquad 2\Delta_{12} = A_1D_2+A_2D_1-B_1C_2-B_2C_1.$$

Now let $A_1A_2 \neq 0$. If $\mathfrak{C}_1$, $\mathfrak{C}_2$ are the circles (γ_1, ρ_1), (γ_2, ρ_2) then by (1.3) and (1.41)

$$(1.52) \qquad \Delta_1 = -A_1^2\rho_1^2, \qquad \Delta_2 = -A_2^2\rho_2^2, \qquad 2\Delta_{12} = A_1A_2(\delta^2-\rho_1^2-\rho_2^2)$$

where

$$\delta = |\gamma_1-\gamma_2|$$

is the distance between their centres.

Further, suppose that $\mathfrak{C}_1$ and $\mathfrak{C}_2$ are both real circles ($A_j \neq 0, \Delta_j < 0, j = 1,2$) and that they have at least one real point in common (cf. Fig. 1). An orientation is fixed on these circles by the sign of the coefficients A_j. Then the angle ω between the two directed circles $\mathfrak{C}_1$, $\mathfrak{C}_2$ will be defined as the angle between the tangents at a common point taken in the direction defined by the orientation. By elementary geometry

$$(1.6) \qquad \delta^2 = \rho_1^2+\rho_2^2\mp 2\rho_1\rho_2 \cos\omega.$$

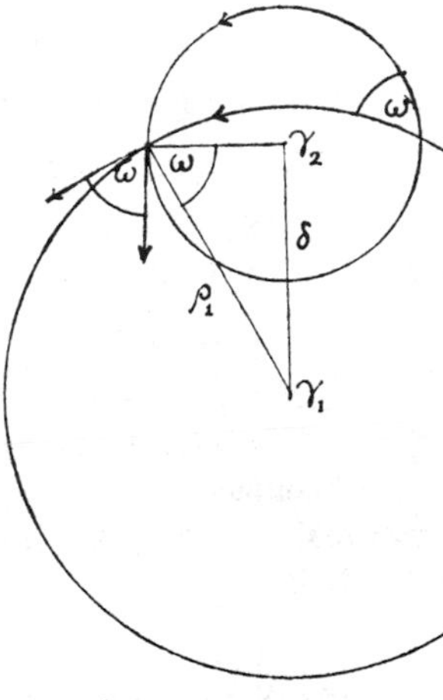

FIG. 1

Hence

$$2\Delta_{12} = \mp 2A_1A_2\rho_1\rho_2 \cos\omega = -2\sqrt{(-\Delta_1)}\sqrt{(-\Delta_2)}\cos\omega,$$

the sign being uniquely determined by the orientations on the two circles $\mathfrak{C}_j$, and

$$(1.61) \qquad \cos\omega = -\frac{\Delta_{12}}{\sqrt{-\Delta_1}\sqrt{-\Delta_2}} = \frac{\Delta_{12}}{\sqrt{\Delta_1}\sqrt{\Delta_2}}.$$

For the real circles $\mathfrak{C}_1$, $\mathfrak{C}_2$ to have points in common it is necessary and sufficient that ω is a real angle:

$$(1.62) \qquad -1 \leqslant \cos\omega \leqslant +1.$$

By (1.6) this is equivalent to

$$\Delta_1\Delta_2 - \Delta_{12}^2 \geqslant 0 \tag{1.63}$$

which together with $\Delta_1 < 0$ is the condition for the quadratic form (1.5) to have only non-positive values for all real λ_1, λ_2.

The equality signs in (1.62) and (1.63) indicate contact or coincidence of the two circles. Suppose that both have positive orientation. If $\cos\omega = +1$, then $\omega = 0$ and therefore $\delta^2 = (\rho_1 - \rho_2)^2$ so that the smaller circle touches the greater one from inside. Similarly if $\cos\omega = -1$, then $\omega = \pi$, $\delta = \rho_1 + \rho_2$, and the two circles touch each other from outside.

Evidently the right-hand term of (1.61) represents a real number for any pair of real or imaginary oriented (directed) circles. We call it their *common invariant* and denote it by $\cos\omega$, even though a real angle ω need not be defined. In the case of two real circles $\mathfrak{C}_1$, $\mathfrak{C}_2$ this will be so if the quadratic form (1.5) may assume positive as well as negative values for different pairs λ_1, λ_2, that is, if

$$\Delta_1\Delta_2 - \Delta_{12}^2 < 0; \tag{1.7}$$

then either

$$\cos\omega > +1,$$

indicating that the smaller circle is entirely contained within the greater one, or

$$\cos\omega < -1,$$

in which case the two circles are lying outside each other. This is readily seen if (1.6) is written in the form

$$(\rho_1 \mp \rho_2)^2 - \delta^2 = 2\rho_1\rho_2(\cos\omega \mp 1). \tag{1.71}$$

Indeed $\cos\omega > 1$ means $|\rho_1 - \rho_2| > \delta$, and $\cos\omega < -1$ similarly $\rho_1 + \rho_2 < \delta$. The same follows by 'continuity' from the preceding discussion of the two different cases of contact.

By (1.6) two real circles $\mathfrak{C}_1$, $\mathfrak{C}_2$ are *orthogonal*, that is, perpendicular to each other, if and only if

$$\delta^2 = \rho_1^2 + \rho_2^2. \tag{1.72}$$

With regard to (1.61) the notion of orthogonality may be extended. A pair of circles, not necessarily real but not point circles ($\Delta_j \neq 0$), is said to be orthogonal if

$$\Delta_{12} = 0, \tag{1.73}$$

which in the case of real circles is equivalent to (1.72).

c. *Pencils of circles.* The two circles $\mathfrak{C}_1$, $\mathfrak{C}_2$ are said to be *generators* of the pencil $\lambda_1\mathfrak{C}_1 + \lambda_2\mathfrak{C}_2$. Any two different circles of the pencil may be taken as generators of the same pencil. Indeed, if with constant μ_j, ν_j

$$\mathfrak{C}_3 = \mu_1\mathfrak{C}_1 + \mu_2\mathfrak{C}_2, \qquad \mathfrak{C}_4 = \nu_1\mathfrak{C}_1 + \nu_2\mathfrak{C}_2$$

are two circles of the pencil, they generate the pencil

$$\lambda_3 \mathfrak{C}_3 + \lambda_4 \mathfrak{C}_4 = (\lambda_3 \mu_1 + \lambda_4 \nu_1)\mathfrak{C}_1 + (\lambda_3 \mu_2 + \lambda_4 \nu_2)\mathfrak{C}_2$$

which coincides with the given pencil provided that for no real λ_3, λ_4, not both equal to zero,

$$\lambda_3 \mu_1 + \lambda_4 \nu_1 = 0$$

$$\lambda_3 \mu_2 + \lambda_4 \nu_2 = 0.$$

This means that $\mu_1 \nu_2 - \mu_2 \nu_1 \neq 0$. For if this determinant vanishes, then μ_1, μ_2 and ν_1, ν_2 are proportional, so that the two circles $\mathfrak{C}_3$, $\mathfrak{C}_4$ coincide and therefore cannot generate a pencil.

There are three different types of pencils:

(i) *The elliptic pencil.* $\Delta_1 \Delta_2 - \Delta_{12}^2 > 0$ or $|\cos \omega| < 1$. The quadratic form (1.5) is negative definite; all circles of the pencil are therefore real circles which pass through two different points z_1, z_2, the common points of $\mathfrak{C}_1$, $\mathfrak{C}_2$.

Actually all circles through these two points belong to the pencil. For any such circle

$$\mathfrak{C} = \begin{pmatrix} A & B \\ C & D \end{pmatrix}$$

has to satisfy the two conditions

(1.8)
$$\begin{cases} \mathfrak{C}(z_1, \bar{z}_1) = A z_1 \bar{z}_1 + B z_1 + C \bar{z}_1 + D = 0 \\ \mathfrak{C}(z_2, \bar{z}_2) = A z_2 \bar{z}_2 + B z_2 + C \bar{z}_2 + D = 0. \end{cases}$$

This is a system of two linear homogeneous equations in A, B, C, D with the matrix

$$\begin{pmatrix} z_1 \bar{z}_1 & z_1 & \bar{z}_1 & 1 \\ z_2 \bar{z}_2 & z_2 & \bar{z}_2 & 1 \end{pmatrix}$$

which has the rank 2 if $z_1 \neq z_2$. Therefore the equations have $4 - 2 = 2$ linearly independent solutions whose components A, B, C, D, however, need not be the elements of hermitian matrices. But we notice that for any given A and B the values of C and D are uniquely defined because the determinant of their coefficients in (1.8), namely $\bar{z}_1 - \bar{z}_2$, is non-zero. Thus let A be real and non-zero. Now both $\mathfrak{C} = \begin{pmatrix} A & B \\ C & D \end{pmatrix}$ and $\bar{\mathfrak{C}}' = \begin{pmatrix} \bar{A} & \bar{C} \\ \bar{B} & \bar{D} \end{pmatrix}$ represent solutions of (1.8). The linear dependence, $\bar{\mathfrak{C}}' = \lambda \mathfrak{C}$, would imply that $\lambda = 1$, and $\mathfrak{C}$ would be hermitian. But since $C = c_1 A + c_2 B$ (c_1, c_2 depending on z_1, z_2), it is clear that a change in the choice of the real A would destroy this condition. Thus $\mathfrak{C}$ and $\bar{\mathfrak{C}}'$ may be assumed to be linearly independent. Then

(1.81)
$$\mathfrak{C}_1 = \mathfrak{C} + \bar{\mathfrak{C}}', \qquad \mathfrak{C}_2 = i(\mathfrak{C} - \bar{\mathfrak{C}}')$$

represent two linearly independent hermitian solutions of (1.8) and

$$\mathfrak{C} = \lambda_1\mathfrak{C}_1 + \lambda_2\mathfrak{C}_2 \qquad \text{with real } \lambda_1, \lambda_2 \tag{1.82}$$

gives the most general hermitian solution; indeed for non-real λ_1 or (and) λ_2 the matrix (1.82) will not be hermitian.

(ii) *The parabolic pencil.* $\Delta_1\Delta_2 - \Delta_{12}^2 = 0$ or $|\cos\omega| = 1$. The quadratic form (1.5) is a complete square with a negative coefficient:

$$-[\sqrt{(-\Delta_1)}\lambda_1 + \sqrt{(-\Delta_2)}\lambda_2]^2;$$

thus the pencil contains exactly one real point circle:

$$\lambda_1 = \sqrt{(-\Delta_2)}, \qquad \lambda_2 = -\sqrt{(-\Delta_1)},$$

situated at the only common point of all circles of the pencil. Here they touch each other ($\omega = 0$ or π). All circles of the pencil are real.

(iii) *The hyperbolic pencil.* $\Delta_1\Delta_2 - \Delta_{12}^2 < 0$ or $|\cos\omega| > 1$. The quadratic form (1.5) is indefinite; thus the pencil contains real circles, imaginary circles, and two different point circles. Two circles of a hyperbolic pencil cannot have a common point.

Pictures of the three types of pencils are shown in Figure 17 and Figure 18.

Finally there is the question of a positive definite (or a non-negative) quadratic form (1.5) which would correspond to a pencil consisting of imaginary circles only (or a pencil where a point circle is the only real circle). It can be shown by algebra (cf. ex. 7) that pencils of this kind do not exist. This will also follow (more easily) by geometrical argument in §4, where a geometrical discussion of the different types of pencils will be given.

EXAMPLES

1. Straight lines, being special cases of circles, are also represented by hermitian matrices $\mathfrak{C}$, viz. those for which $A = 0$.

(a) Express the non-zero coefficients of the hermitian matrix of a straight line, viz. B, C, D, by means of the distance p of the line from the origin of the coordinate system, and the angle θ which the distance vector makes with the x-axis.

(b) Verify the general formula (1.61) if $\mathfrak{C}_1$ and $\mathfrak{C}_2$ are two straight lines.

2. What is the relative position of the circles $\mathfrak{C}$ and $\mathfrak{C}' = \bar{\mathfrak{C}}$?

3. Let $\mathfrak{C}$ be any circle. Find the matrix $\mathfrak{C}_0$ of the circle obtained from $\mathfrak{C}$ by a translation $\tilde{z} = z + b$ such that 0 is the centre of the new circle $\mathfrak{C}_0$.

4. (a) Show that every pencil of non-concentric circles contains at least one straight line, and that it consists of straight lines only if it contains more than one straight line.

(b) If the pencil contains one straight line only, describe the position of this line within the pencil. (In the case of a hyperbolic pencil take the two point circles as generators.)

5. Is contact possible between a real and an imaginary circle?

6. Calculate the common invariant $\cos\omega$ for two concentric circles $(0, \rho_1)$ and $(0, \rho_2)$.

7. Let $\mathfrak{C}_1$, $\mathfrak{C}_2$ be two imaginary circles, represented by two positive definite hermitian matrices. The pencil $\lambda_1\mathfrak{C}_1+\lambda_2\mathfrak{C}_2$ then always contains real circles. (Notice that $A_2\mathfrak{C}_1-A_1\mathfrak{C}_2$ is a straight line and *eo ipso* real.)

Remark. This fact involves a rather general inequality between complex numbers. Let B_1, B_2 be two complex numbers, A_1, D_1, A_2, D_2 four positive numbers and

$$A_1D_1 > |B_1|^2, \qquad A_2D_2 > |B_2|^2;$$

then

$$\Delta_1\Delta_2-\Delta_{12}{}^2 = -\tfrac{1}{4}(A_1D_2-A_2D_1)^2+$$
$$+(\operatorname{Im} B_1\bar{B}_2)^2+\operatorname{Re}\{(A_1\bar{B}_2-A_2\bar{B}_1)(D_2B_1-D_1B_2)\} < 0.$$

For a geometrical verification cf. § 4, **a**.

8. Prove that the two circles $\mathfrak{C}_1$, $\mathfrak{C}_2$ are orthogonal if and only if

$$\operatorname{tr}(\mathfrak{C}_1\mathfrak{C}_2^{-1}) = 0. \tag{1.9}$$

(By tr A we denote the trace, that is, the sum of the diagonal elements, of the matrix A.)

9. The equation

$$z = \frac{at+b}{ct+d} \qquad -\infty < t < \infty$$

is a parametric representation of a real circle. Find the hermitian matrix $\mathfrak{C}$ of this circle. The circle will become a straight line if and only if $c = 0$ or if d/c is real.

10. A point circle $\mathfrak{C} = \begin{pmatrix} A & B \\ C & D \end{pmatrix}$ is orthogonal to a circle $\mathfrak{C}_0 = \begin{pmatrix} A_0 & B_0 \\ C_0 & D_0 \end{pmatrix}$ if and only if its only point (and centre) is a point of the circumference of $\mathfrak{C}_0$. Assume $A = 1$; then $D = BC$ and $z = -C$ is *the* point of $\mathfrak{C}$. The orthogonality condition

$$A_0D+D_0A-B_0C-C_0B = A_0z\bar{z}+B_0z+C_0\bar{z}+D_0 = 0$$

expresses the fact that this is a point of the circle $\mathfrak{C}_0$.

11. Every circle $\mathfrak{C}$ orthogonal to an imaginary circle $\mathfrak{C}_0$ is real. Assume $A_0 = 1$; then $D_0 > B_0\bar{B}_0$. Since (by orthogonality) $D = B_0C+C_0B-D_0A$, it follows that

$$AD-BC = A(B_0C+C_0B)-D_0A^2-BC < A(B_0\bar{B}+\bar{B}_0B)-A^2B_0\bar{B}_0-B\bar{B}$$
$$= -|B-AB_0|^2 \leqslant 0.$$

12. In order to determine a circle $\mathfrak{C}$ within a pencil $\mathfrak{C}_1+\lambda\mathfrak{C}_2$ it is useful to study the radius $\rho = \rho(\lambda)$ as a function of λ. By (1.41)

$$\rho(\lambda) = \frac{\sqrt{(-|\mathfrak{C}_1+\lambda\mathfrak{C}_2|)}}{A_1+\lambda A_2}.$$

The differential coefficient of this function is

$$\rho'(\lambda) = \frac{A_2\Delta_1-A_1\Delta_{12}+(A_2\Delta_{12}-A_1\Delta_2)\lambda}{(A_1+A_2\lambda)^2\sqrt{[-(\Delta_1+2\Delta_{12}\lambda+\Delta_2\lambda^2)]}}.$$

The sign of

$$\rho'(0) = \frac{A_2}{A_1}\frac{\rho_2^2-\delta^2-\rho_1^2}{2\rho_1}$$

indicates whether for circles $\mathfrak{C}$ near $\mathfrak{C}_1$ the parameter λ increases or decreases and therefore is positive or negative if the radius ρ increases from ρ_1 on.

§ 2. The inversion

a. *Definition.* The definition of the inversion is based on the following theorem.

Theorem A. *Let $\mathfrak{C}_0$ be a circle (not a point circle) and z a point, neither on the circle $\mathfrak{C}_0$ nor at its centre. Then there is one and only one point z^* different from z which is common to all circles through z that are orthogonal to $\mathfrak{C}_0$.*

The point z^* is called the *inverse* of z with respect to the 'fundamental circle' $\mathfrak{C}_0$, which may be real, imaginary, or a straight line. The transformation carrying z into z^* (or leading from z to z^*) is called the *inversion* with respect to $\mathfrak{C}_0$. The inversion is an *involutory* transformation, that is, a transformation which coincides with its own inverse; in fact, according to Theorem A the point z is the inverse of z^*.

If $\mathfrak{C}_0$ is a straight line, the point z^* is symmetric to z with respect to $\mathfrak{C}_0$. If $\mathfrak{C}_0$ is a real circle (γ_0, ρ_0), the straight line through γ_0 and z is an orthogonal circle to $\mathfrak{C}_0$. Hence z^* must lie on this line and will be determined by one more orthogonal circle. For illustration let $\mathfrak{C}_0$ be the unit circle $z\bar{z}-1 = 0$, $\mathfrak{C}_0 = \begin{pmatrix}1 & 0\\ 0 & -1\end{pmatrix}$, and let $z = x$ be real ($x \neq 0, \neq \pm 1$). Then $z^* = 1/x$. Indeed the centres of all circles $\mathfrak{C}$ through x and $1/x$ have the abscissa $\alpha = \frac{1}{2}(x+1/x)$. If ρ is the radius of $\mathfrak{C}$, the ordinate β of its centre will be found from the relation $\beta^2 = \rho^2-(\alpha-x)^2$ (cf. Figure 2). Thus

$$\delta^2 = \alpha^2+\beta^2 = \rho^2+1$$

which according to (1.72) means that $\mathfrak{C}_0$ and $\mathfrak{C}$ are orthogonal whatever the radius

$$\rho \geqslant \frac{1}{2}\left|x-\frac{1}{x}\right|.$$

Conversely, if a circle $\mathfrak{C}$, orthogonal to $\mathfrak{C}_0$, passes through the point x, then it also passes through the point $1/x$. For the orthogonality condition is in this

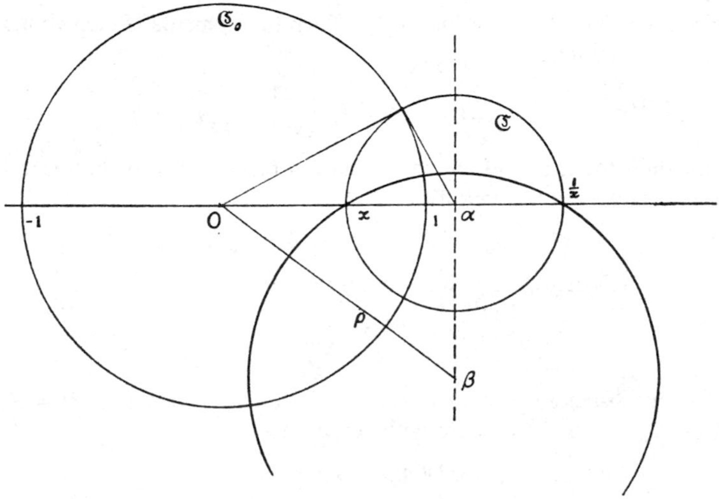

FIG. 2

case $A-D = 0$, and the equation $Ax^2+(B+C)x+A = 0$, expressing that x is a point of $\mathfrak{C}$, has $1/x$ as its second root.

Proof of Theorem A. Let $\mathfrak{C}_0 = \begin{pmatrix} A_0 & B_0 \\ C_0 & D_0 \end{pmatrix}$ be the hermitian matrix of the fundamental circle and thus $z \neq -C_0/A_0$. We have to determine all circles $\mathfrak{C} = \begin{pmatrix} A & B \\ C & D \end{pmatrix}$ passing through z and orthogonal to $\mathfrak{C}_0$, and to show that they all pass through a certain point z^*. Thus we obtain the following three linear homogeneous equations in the four unknowns A, B, C, D:

$$(2.1)\qquad \begin{cases} \mathfrak{C}(z, \bar{z}) = Az\bar{z}+Bz+C\bar{z}+D = 0: & z \text{ on } \mathfrak{C} \\ \Delta_{01} = AD_0-BC_0-CB_0+DA_0 = 0: & \mathfrak{C}\perp\mathfrak{C}_0 \\ \mathfrak{C}(z^*, \overline{z^*}) = Az^*\overline{z^*}+Bz^*+C\overline{z^*}+D = 0: & z^* \text{ on } \mathfrak{C}. \end{cases}$$

They always have a non-zero solution. However a single linearly independent hermitian solution $\mathfrak{C}$ would correspond to a single orthogonal circle $\mathfrak{C}$ through

z and z^*. In order to obtain a two-parameter family of solutions we have to find z^* such that the matrix of the three equations (2.1), viz.

$$\text{(2.11)} \qquad \begin{pmatrix} z\bar{z} & z & \bar{z} & 1 \\ D_0 & -C_0 & -B_0 & A_0 \\ z^*\overline{z^*} & z^* & \overline{z^*} & 1 \end{pmatrix}$$

has the rank 2.

Let us assume that z^* has this property. Then the elements of the two matrices (in general non-hermitian)

$$\text{(2.12)} \qquad \mathfrak{C}^{(1)} = \begin{pmatrix} 1 & -\bar{z} \\ -z^* & z^*\bar{z} \end{pmatrix}, \qquad \mathfrak{C}^{(2)} = \begin{pmatrix} 1 & -\overline{z^*} \\ -z & z\overline{z^*} \end{pmatrix} = \overline{\mathfrak{C}^{(1)\prime}}$$

satisfy not only the first and the last equation (2.1), but also the second one. This gives in each case the condition

$$\text{(2.2)} \qquad A_0 z^*\bar{z} + B_0 z^* + C_0\bar{z} + D_0 = \mathfrak{C}_0(z^*, \bar{z}) = 0$$

by which z^* is uniquely defined as a function of z:

$$\text{(2.21)} \qquad z^* = -\frac{C_0\bar{z}+D_0}{A_0\bar{z}+B_0}.$$

The previous discussion remains valid if z lies on $\mathfrak{C}_0$. Then $\mathfrak{C}_0(z, \bar{z}) = A_0 z\bar{z} + B_0 z + C_0\bar{z} + D_0 = 0$ and therefore with respect to (2.2)

$$(z-z^*)(A_0\bar{z}+B_0) = 0.$$

Since by hypothesis $A_0\bar{z}+B_0 = \overline{A_0 z + C_0} \neq 0$, it follows that $z^* = z$. Hence every point of the (circumference of the) fundamental circle $\mathfrak{C}_0$ is its own inverse. These points are called the *fixed* (or *invariant*) points of the inversion.

The circle $\mathfrak{C}_0$ was assumed not to be a point circle; hence the determinant $|\mathfrak{C}_0| \neq 0$. Therefore the transformation (2.21) representing the inversion can be inverted (by solving with respect to z) and indeed the inverse transformation coincides with the original.

If the determinant $|\mathfrak{C}_0| = 0$ (and $A_0 \neq 0$) so that $\mathfrak{C}_0$ is a point circle, then

$$D_0 = \frac{B_0 C_0}{A_0}.$$

Therefore by (2.21)

$$z^* = -\frac{C_0}{A_0} = \text{const.}$$

is independent of z. This means that the transformation carries every point z into one and the same point z^*; thus it cannot be invertible.

The inversion (2.21) is also called *symmetry* with respect to the fundamental circle $\mathfrak{C}_0$, and z and z^* are said to be symmetric with respect to $\mathfrak{C}_0$. Equation

(2.2) defines symmetry between two points, z, z^*; from the equation $\mathfrak{C}_0(z, \bar{z}) = 0$ of the fundamental circle $\mathfrak{C}_0$ it is obtained by formally replacing z by z^* and leaving $\bar{z}$ unchanged. If we assume the equation of the fundamental circle in the form $(z-\gamma_0)(\bar{z}-\bar{\gamma}_0) = \rho_0^2$, we obtain by the same formal procedure

(2.22)
$$z^* = \gamma_0 + \frac{\rho_0^2}{\bar{z}-\bar{\gamma}_0}.$$

If $\mathfrak{C}_0$ is real ($\rho^2 > 0$) the inversion is called *hyperbolic*. With the real circle $\mathfrak{C}_0 = (\gamma_0, \rho_0)$ one associates the imaginary circle $(\gamma_0, i\rho_0)$. With respect to this circle the inverse point z_* of z is given by

(2.3)
$$z_* = \gamma_0 - \frac{\rho_0^2}{\bar{z}-\bar{\gamma}_0}.$$

This is the *elliptic* inversion.

The point z_* may also be obtained from z by following up the hyperbolic inversion (2.22) (with respect to the associated real circle $\mathfrak{C}_0$) by the symmetry with respect to the centre γ_0:

(2.31)
$$z_* = 2\gamma_0 - z^*.$$

b. *Simple properties of the inversion.* Every point in the plane, except the centre $z = \gamma_0 = -C_0/A_0$ of the fundamental circle, is carried by the inversion into a definite image point z^*, given by (2.21) or (2.22). Let $z^* = f(z)$ be this relation. In order to make the inversion a one-one correspondence of the whole plane onto itself, it is necessary to complete the ordinary complex number plane by adding to it one further element which is called the *point at infinity* and represented by the symbol ∞. Thus we complete the definition of the inversion by putting

(2.4)
$$f(\gamma_0) = \infty.$$

Because the inversion is involutory, that is, $f(z^*) = z$, it will be natural to put

(2.41)
$$f(\infty) = \gamma_0,$$

after which the inversion $z^* = f(z)$ is a one–one correspondence of the 'completed plane' onto itself.

From (2.21) or (2.22) one derives

$$\lim_{z\to\gamma_0} f(z) = \infty, \qquad \lim_{z\to\infty} f(z) = \gamma_0$$

which with regard to (2.4) and (2.41) expresses the continuity of the inversion at the centre γ_0 and at ∞ respectively.

Definition. A transformation $Z = \phi(z)$ of the z-plane onto the Z-plane (which may be considered as coincident with the z-plane, with coincident coordinate systems and equal scales on the axes) is said to be isogonal at the point z_0, if it carries any two curves meeting each other at z_0, making the angle[2] ω, into two curves making the angle $\Omega = \pm\omega$ at $Z_0 = \phi(z_0)$. The transformation is said to be conformal at z_0, if $\Omega = \omega$.

We shall now prove the following two theorems.

Theorem B. *Every inversion carries circles into circles, real circles (including straight lines) into real circles, imaginary circles into imaginary circles.*

Theorem C. *The inversion is an isogonal transformation changing the angle between two curves into the negative angle:*

$$\omega^* = -\omega.$$

In virtue of the concluding remark in subsection **a** it will be sufficient to prove these statements for hyperbolic inversion only. It is, indeed, evident, that the symmetry (2.31) which produces an elliptic inversion from a certain hyperbolic inversion (i) carries circles into circles, (ii) is a conformal transformation. Thus if we follow up a circle-preserving isogonal mapping by such a symmetry (2.31), another circle-preserving isogonal mapping will result.

Proof of Theorem B. We assume that the fundamental circle $\mathfrak{C}_0$ of the inversion is a proper real circle ($|\mathfrak{C}_0| < 0$, $A_0 \neq 0$). For the case of a straight line cf. ex. 3. Because of the geometrical definition of the inversion (based upon Theorem A) its geometrical properties do not depend on the position of $\mathfrak{C}_0$ in the plane nor on the magnitude of the radius ρ_0. Hence we may choose $\gamma_0 = 0$, $\rho_0 = 1$. By (2.22) the inversion is then given by $z^* = 1/\bar{z}$. Let $\mathfrak{C}$ be any circle. Its image by the inversion is then obtained by substituting $z = 1/\overline{z^*}$ in (1.2). After multiplication by the positive factor $z^*\overline{z^*}$ the equation will be

$$z^*\overline{z^*}\mathfrak{C}\left(\frac{1}{\overline{z^*}}, \frac{1}{z^*}\right) = Dz^*\overline{z^*} + Bz^* + C\overline{z^*} + A = 0. \tag{2.5}$$

This is the equation of the image circle $\mathfrak{C}^* = \begin{pmatrix} D & B \\ C & A \end{pmatrix}$. Its discriminant is $\Delta^* = \Delta$; thus $\mathfrak{C}^*$ is real if $\mathfrak{C}$ is real and imaginary if $\mathfrak{C}$ is imaginary.

Remark. A circle $\mathfrak{C}$ passing through the centre γ_0 of the fundamental circle ($\gamma_0 = 0$ in the proof) will be carried by the inversion into a circle passing through the point ∞, that is, a straight line. In fact no other real circle passes through ∞. Algebraically (in the terms of the proof): If the circle $\mathfrak{C}$ passes through 0, then $D = 0$ and therefore $\mathfrak{C}^*$ is a straight line.

Proof of Theorem C. For two circles $\mathfrak{C}_1, \mathfrak{C}_2$ we find the images $\mathfrak{C}_1^*, \mathfrak{C}_2^*$ according to (2.5). Then $\Delta_1^* = \Delta_1$, $\Delta_2^* = \Delta_2$, and also $\Delta_{12}^* = \Delta_{12}$; therefore by (1.61)

$$\cos\omega^* = \cos\omega \tag{2.6}$$

[2] The angle of two curves at a point of intersection is, by definition, the angle of their tangents at this point.

without change of sign, because inversion turns the orientation of any real circle into the opposite. This can easily be inferred from the geometrical construction indicated in Figure 3. Thus $\omega^* = \pm\omega$. By the same argument, however, one concludes that $\omega^* = -\omega$.

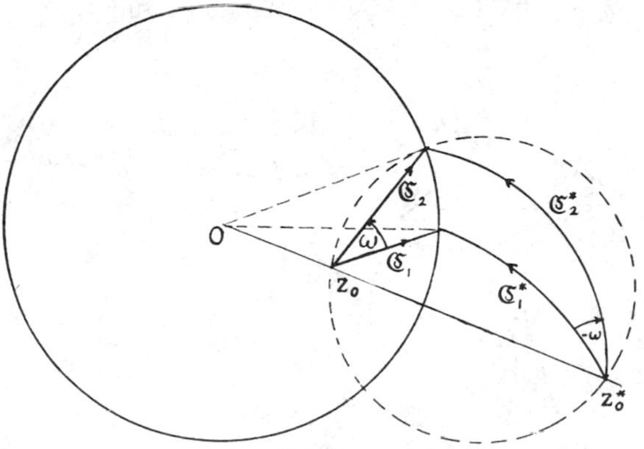

FIG. 3

Remark. The relation (2.6) is an algebraic identity, valid for any two circles $\mathfrak{C}_1$, $\mathfrak{C}_2$, even if they have no point in common. It justifies the name of 'common invariant' for cos ω. There is a simple geometrical interpretation only if $\mathfrak{C}_1$, $\mathfrak{C}_2$ are real and have a common point.

EXAMPLES

1. If the point z moves on a real circle $\mathfrak{C}$ orthogonal to the (real or imaginary) circle $\mathfrak{C}_0$, then, by Theorem A, the inverse z^* of z with respect to $\mathfrak{C}_0$ moves on the same circle $\mathfrak{C}$. Conversely if z and z^* both move on a circle $\mathfrak{C}$, then $\mathfrak{C}$ is orthogonal to $\mathfrak{C}_0$. Thus: *a real circle $\mathfrak{C} \neq \mathfrak{C}_0$ is invariant under inversion with respect to $\mathfrak{C}_0$ if and only if $\mathfrak{C}$ and $\mathfrak{C}_0$ are orthogonal.* If $\mathfrak{C}_0$ is real, the inversion with respect to $\mathfrak{C}_0$ maps the interior of $\mathfrak{C}$ onto itself (cf. § 6, ex. 13).

2. For a given point $z \neq 0$ construct geometrically the points $1/\bar{z}$ and $1/z$.

3. Study the inversion (symmetry or reflection) with respect to a straight line as fundamental circle. It maps proper circles onto proper circles and straight lines onto straight lines.

4. For two point circles $\mathfrak{C}_1$, $\mathfrak{C}_2$ show that their centres γ_1, γ_2 are mutually inverse with respect to any circle of the pencil

$$\lambda_1\mathfrak{C}_1 + \lambda_2\mathfrak{C}_2 \qquad (\lambda_1, \lambda_2 \text{ real}).$$

5. *Solution of the equations* (2.1). Suppose that z is not a point of the fundamental circle $\mathfrak{C}_0$. Assuming the condition (2.2) the elements of the two matrices $\mathfrak{C}^{(1)}$, $\mathfrak{C}^{(2)}$ of (2.12) constitute a basis for the solution of (2.1). By suitable linear combination of $\mathfrak{C}^{(1)}$, $\mathfrak{C}^{(2)}$ a hermitian basis can be obtained:

$$\mathfrak{C}_1 = \tfrac{1}{2}(\mathfrak{C}^{(1)} + \mathfrak{C}^{(2)}), \qquad \mathfrak{C}_2 = \tfrac{1}{2}i\,(\mathfrak{C}^{(1)} - \mathfrak{C}^{(2)}) \qquad \text{[cf. § 1, c, (i)].}$$

In fact since $\overline{\mathfrak{C}^{(1)\prime}} = \mathfrak{C}^{(2)}$, it follows that $\overline{\mathfrak{C}_1}' = \mathfrak{C}_1$, $\overline{\mathfrak{C}_2}' = \mathfrak{C}_2$. Thus $\mathfrak{C}_1$ and $\mathfrak{C}_2$ are the hermitian matrices of two different circles passing through z and z^*:

$$\mathfrak{C}_1(z, \bar{z}) = 0, \qquad \mathfrak{C}_1(z^*, \overline{z^*}) = 0, \qquad \mathfrak{C}_2(z, \bar{z}) = 0, \qquad \mathfrak{C}_2(z^*, \overline{z^*}) = 0.$$

They generate the pencil $\lambda_1\mathfrak{C}_1 + \lambda_2\mathfrak{C}_2$ of all circles passing through z and orthogonal to $\mathfrak{C}_0$.

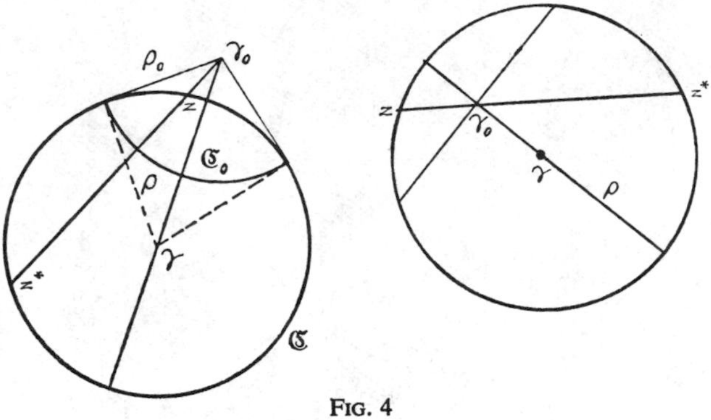

FIG. 4

6. The *polar* of a point $z_0 \neq 0$, with respect to the circle $\mathfrak{C}_+ = (0, \rho)$ (or $\mathfrak{C}_- = (0, i\rho)$), is represented by the equation

$$\bar{z}_0 z + z_0 \bar{z} = 2\rho^2 \text{ (or } = -2\rho^2).$$

Show that the polar of z_0 passes through the inverse point z_0^* of z_0 with respect to this circle. It is perpendicular to the radius vector $\overrightarrow{0z_0}$. If z_0 is a point of the circle, then the polar is the tangent to the circle at z_0. Let z_0 be outside the real circle $\mathfrak{C}_+$; its polar intersects the circle at the two points z_1, z_2. Then the tangents to the circle at z_1 and z_2 meet each other at z_0.

7. Let γ_0 be a point not on the circle $\mathfrak{C}$. A circle $\mathfrak{C}_0 = (\gamma_0, \rho_0)$ can then be found which is orthogonal to $\mathfrak{C}$. If $\mathfrak{C}$ is real, the circle $\mathfrak{C}_0$ will be real if γ_0 lies outside $\mathfrak{C}$, imaginary if γ_0 lies inside $\mathfrak{C}$; if $\mathfrak{C}$ is imaginary, the circle $\mathfrak{C}_0$ will always be real (cf. § 1, ex. 11). Let $\mathfrak{C} = (\gamma, \rho)$, then by (1.72)

$$\rho_0^2 = |\gamma - \gamma_0|^2 - \rho^2.$$

This constant is called the *power* of the point γ_0 with respect to the circle $\mathfrak{C}$. It is positive if γ_0 lies outside $\mathfrak{C}$, negative if γ_0 lies inside $\mathfrak{C}$.

In either case two points z, z^* on $\mathfrak{C}$ which are situated on one and the same line through γ_0 are symmetric with respect to $\mathfrak{C}_0$. Hence

$$(\bar{z} - \bar{\gamma}_0)(z^* - \gamma_0) = \rho_0^2 = \text{const.} \qquad \text{(cf. Figure 4).}$$

8. *Fixed points.* The point z_0 is called a fixed or invariant point of the transformation $Z = f(z)$ if it satisfies the equation $f(z_0) = z_0$. A hyperbolic inversion

interchanges the interior and the exterior of each radius of the fundamental circle; it has therefore the points of this circle and no others as fixed points. An elliptic inversion (imaginary fundamental circle) possesses no fixed points. A circle being entirely defined by three of its points, it is clear that any hyperbolic inversion is defined by three of its fixed points.

9. Let $\mathfrak{C}_0$ be a fundamental circle for an inversion and let $\mathfrak{C}_1$ be another given circle (real or imaginary). The circle $\mathfrak{C}_1^*$, symmetric to $\mathfrak{C}_1$ with respect to $\mathfrak{C}_0$, is a circle of the pencil generated by $\mathfrak{C}_0$ and $\mathfrak{C}_1$. Thus

$$\mathfrak{C}_1^* = \lambda_0\mathfrak{C}_0+\lambda_1\mathfrak{C}_1 \tag{2.7}$$

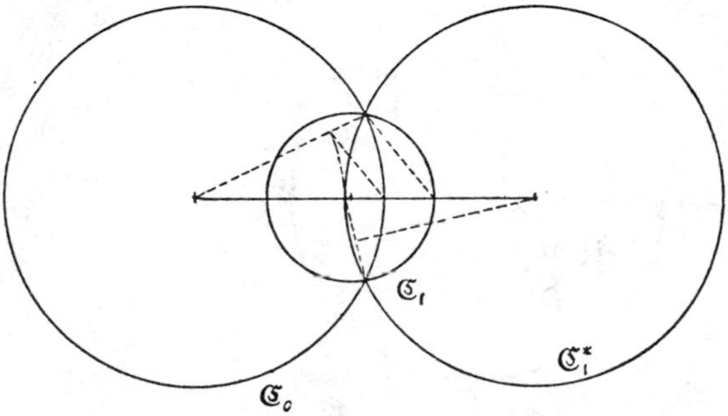

FIG. 5, a

with real λ_0, λ_1, not both zero. This is geometrically evident when $\mathfrak{C}_0$ and $\mathfrak{C}_1$ have one or two real points in common. In this case the pencil is parabolic or elliptic. If there is no common point the pencil is hyperbolic. Corresponding points, z on $\mathfrak{C}_1$ and z^* on $\mathfrak{C}_1^*$, lie on the circle through z which is orthogonal to $\mathfrak{C}_1$ and to $\mathfrak{C}_0$.

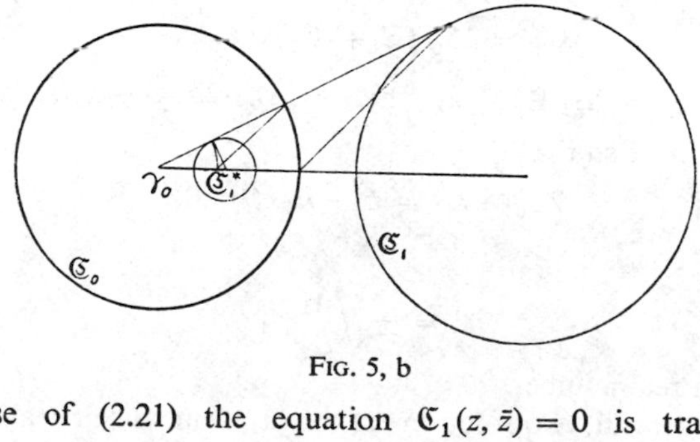

FIG. 5, b

Making use of (2.21) the equation $\mathfrak{C}_1(z, \bar{z}) = 0$ is transformed into $\mathfrak{C}_1^*(z^*, \overline{z^*}) = 0$ which must be the same as $(\lambda_0\mathfrak{C}_0+\lambda_1\mathfrak{C}_1)(z^*, \overline{z^*}) = 0$. Hence

$$\begin{cases} \lambda_0 = 2\Delta_{01} = A_0D_1+A_1D_0-B_0C_1-B_1C_0 \\ \lambda_1 = -\Delta_0 = B_0C_0-A_0D_0. \end{cases} \tag{2.71}$$

In the case of real circles $\mathfrak{C}_0$, $\mathfrak{C}_1$ the geometrical construction of $\mathfrak{C}_1^*$ is shown in Figure 5, a–c, for three different situations.

10. By means of (2.71) one can solve the converse problem: for two given circles $\mathfrak{C}_1$, $\mathfrak{C}_2$ (both real or both imaginary) find the fundamental circle $\mathfrak{C}_0$ such that $\mathfrak{C}_2 = \mathfrak{C}_1^*$, that is, the inverse (symmetric) circle of $\mathfrak{C}_1$ with respect to $\mathfrak{C}_0$.

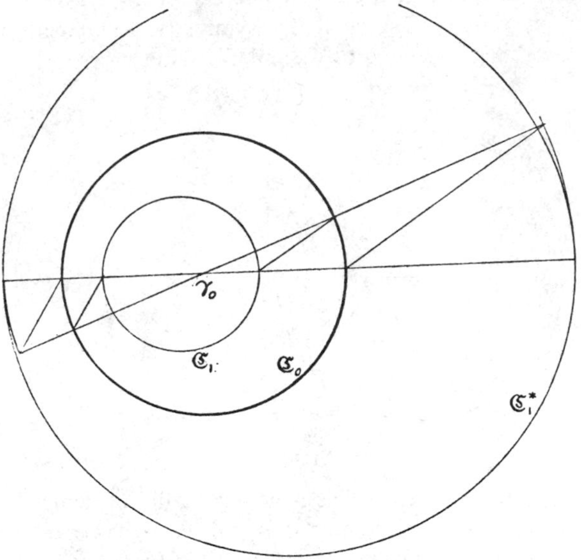

FIG. 5, c

Evidently $\mathfrak{C}_0$ must belong to the pencil generated by $\mathfrak{C}_1$ and $\mathfrak{C}_2$. Thus by (2.71)

$$\mathfrak{C}_1^* = 2\Delta_{01}(\mathfrak{C}_1+\lambda\mathfrak{C}_2)-\Delta_0\mathfrak{C}_1$$

where

$$2\Delta_{01} = 2\Delta_1+2\Delta_{12}\lambda, \qquad \Delta_0 = \Delta_1+2\Delta_{12}\lambda+\Delta_2\lambda^2.$$

Now we determine λ so that

$$2\Delta_{01}-\Delta_0 = \Delta_1-\Delta_2\lambda^2 = 0,$$

that is,

$$\lambda = \pm\sqrt{\frac{\Delta_1}{\Delta_2}}. \tag{2.72}$$

This is always a real number.

There are, in general, two solutions of the problem. Their reality depends on the sign of the discriminant

$$|\mathfrak{C}_0| = \left|\mathfrak{C}_1 \pm \sqrt{\left(\frac{\Delta_1}{\Delta_2}\right)}\mathfrak{C}_2\right| = 2\Delta_1\left(1 \pm \frac{\Delta_{12}}{\sqrt{\Delta_1}\sqrt{\Delta_2}}\right).$$

1. $\rho_1 + \rho_2 > \delta : \Delta_0^{(1)} < 0$

$|\rho_1 - \rho_2| > \delta : \Delta_0^{(2)} > 0$

2. $\rho_1 + \rho_2 > \delta : \Delta_0^{(1)} < 0$

$|\rho_1 - \rho_2| = \delta : \Delta_0^{(2)} = 0$

3. $\rho_1 + \rho_2 > \delta : \Delta_0^{(1)} < 0$

$|\rho_1 - \rho_2| < \delta : \Delta_0^{(2)} < 0$

5. $\rho_1 + \rho_2 < \delta : \Delta_0^{(1)} > 0$

$|\rho_1 - \rho_2| < \delta : \Delta_0^{(2)} < 0$

4. $\rho_1 + \rho_2 = \delta : \Delta_0^{(1)} = 0$

$|\rho_1 - \rho_2| < \delta : \Delta_0^{(2)} < 0$

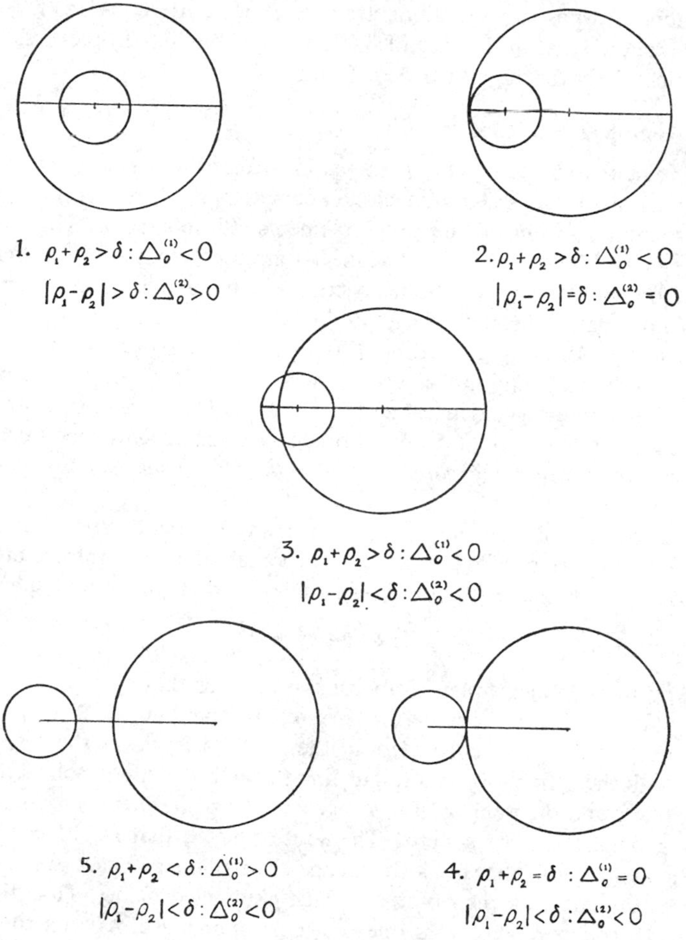

Fig. 6

In the case of real circles $\mathfrak{C}_1 = (\gamma_1, \rho_1)$, $\mathfrak{C}_2 = (\gamma_2, \rho_2)$ one has thus from (1.52) the two values

$$|\mathfrak{C}_0| = \begin{cases} \Delta_0^{(1)} = A_1^2 \dfrac{\rho_1}{\rho_2}[\delta^2 - (\rho_1 + \rho_2)^2] \\[2mm] \Delta_0^{(2)} = A_1^2 \dfrac{\rho_1}{\rho_2}[(\rho_1 - \rho_2)^2 - \delta^2]. \end{cases}$$

There is always one real circle $\mathfrak{C}_0$ since one of the values $\Delta_0^{(1)}$, $\Delta_0^{(2)}$ will be negative. Both are negative if and only if $\mathfrak{C}_1$ and $\mathfrak{C}_2$ have two different points in common. All possible situations are shown in Figure 6. Also cf. § 4, ex. 8 where it is shown that in the case of intersecting circles $\mathfrak{C}_1$, $\mathfrak{C}_2$ the circles $\mathfrak{C}_0$ are those which bisect the angles between $\mathfrak{C}_1$ and $\mathfrak{C}_2$.

§ 3. Stereographic projection

a. *Definition.* The completed plane as introduced in § 2, **b**, often turns out to be unsatisfactory as a stage for circle geometry and its applications because of the isolated position of one of its elements, the point ∞. Therefore it is desirable to replace the completed plane by another geometrical carrier of the complex numbers where no point occupies a distinguished position. Geometrical imagination locating the point ∞ at a 'very large' distance from the observer in the plane suggests taking this element to fill an existing hole or gap in the ordinary plane and in this way closing it at infinity. By this process the plane is turned into a geometrical surface of the nature of a sphere. It remains to give an accurate description of this process by establishing a certain *one–one correspondence between the points of the completed plane and the points of a sphere in space.*

Such a correspondence can be set up in many different ways. One which is of particular importance for our present purposes is the stereographic projection. In a cartesian coordinate system (ξ, η, ζ) in space we take the unit sphere

$$\xi^2+\eta^2+\zeta^2 = 1 \tag{3.1}$$

and we put the complex z-plane into the ξ, η-plane so that for a point $z = x+iy$ in this plane $\xi = x$, $\eta = y$, $\zeta = 0$. The stereographic image $\mathbf{P}$ (ξ, η, ζ) on the sphere (3.1) of the point $z = x+iy$ in the plane is found as the second intersection with the sphere of the straight line through the south pole $\mathbf{S}$ $(0, 0, -1)$ of the sphere and the point z. For every z in the plane there is a unique corresponding point $\mathbf{P}$ on the sphere. The south pole $\mathbf{S}$, that is, the centre of the projection, does not appear as the image of a point z; hence we are free to take $\mathbf{S}$ as the image of the point ∞ of the completed plane. Thus the stereographic projection represents a one–one correspondence between the points z of the completed plane and the points $\mathbf{P}$ of the sphere.

Now we establish the equations of the stereographic projection. Let $\mathbf{P}_1(\xi_1, \eta_1, \zeta_1)$, $\mathbf{P}_2(\xi_2, \eta_2, \zeta_2)$ be two points in space. Any point $\mathbf{Q}(\xi, \eta, \zeta)$ of the line $\mathbf{P}_1\mathbf{P}_2$ is then given in the form

$$\mathbf{Q} = (1-\lambda)\mathbf{P}_1+\lambda\mathbf{P}_2 \qquad \text{(real parameter } \lambda\text{)},$$

that is,

$$\begin{cases} \xi = (1-\lambda)\xi_1+\lambda\xi_2 \\ \eta = (1-\lambda)\eta_1+\lambda\eta_2 \\ \zeta = (1-\lambda)\zeta_1+\lambda\zeta_2. \end{cases} \tag{3.2}$$

In order to find the stereographic image **P** of the point $z = x+iy$, let

$$\mathbf{P}_1 = \mathbf{S}(0, 0, -1), \qquad \mathbf{P}_2 = (x, y, 0).$$

Then $\mathbf{P} = \mathbf{Q}$ and by (3.2)

$$\xi = \lambda x, \qquad \eta = \lambda y, \qquad \zeta = -(1-\lambda).$$

The value of λ is determined by the condition (3.1):

$$\xi^2+\eta^2+\zeta^2 = \lambda^2(x^2+y^2)+1-2\lambda+\lambda^2 = 1,$$

whence (excluding the value $\lambda = 0$ which corresponds to the point **S**):

$$\lambda = \frac{2}{1+x^2+y^2} = \frac{2}{1+z\bar{z}}.$$

Thus the coordinates of $\mathbf{P}(\xi, \eta, \zeta)$ are given by

$$\xi+i\eta = \frac{2z}{1+z\bar{z}}, \qquad \zeta = \frac{1-z\bar{z}}{1+z\bar{z}}. \tag{3.3}$$

To find the point z for a given **P** on the sphere, we apply again (3.2) taking $\mathbf{P}_1 = \mathbf{S}$, $\mathbf{P}_2 = \mathbf{P}$ and $\mathbf{Q} = (x, y, 0)$; thus

$$x = \lambda\xi, \qquad y = \lambda\eta, \qquad 0 = -(1-\lambda)+\lambda\zeta.$$

Hence $\lambda = 1/(1+\zeta)$ and therefore

$$z = \frac{\xi+i\eta}{1+\zeta}. \tag{3.31}$$

To every point $\mathbf{P} \neq \mathbf{S}$ on the sphere (that is, $\zeta \neq -1$) corresponds, by (3.31), a point z of the complex number plane. As **P** approaches the point **S**, $\zeta \to -1$, and since

$$|z|^2 = \frac{\xi^2+\eta^2}{(1+\zeta)^2} = \frac{1-\zeta^2}{(1+\zeta)^2} = \frac{1-\zeta}{1+\zeta},$$

it follows that $|z| \to \infty$, that is, $z \to \infty$. The stereographic projection of the sphere into the completed plane is therefore continuous everywhere on the sphere, including the point **S**.

To every z corresponds by (3.3) a point **P** on the sphere (3.1). If now $z \to \infty$, that is,

$$\frac{1}{z} = \frac{1+\zeta}{\xi+i\eta} \to 0,$$

then $\zeta \to -1$ and at the same time by (3.1) $\xi^2+\eta^2 \to 0$, that is, $\xi \to 0$ and $\eta \to 0$, which means that **P** approaches the point **S**. Hence also the stereographic projection of the plane to the sphere is a continuous mapping, in particular continuous at the point ∞.

A mapping of a surface onto another surface, which is continuous, invertible, and whose inverse mapping is likewise continuous, is called a *topological mapping*. Two surfaces which may be carried into one another by a topological mapping, are said to be *topologically equivalent*. Hence

Theorem A. *The stereographic projection is a topological mapping of the sphere onto the completed plane; these two surfaces are therefore topologically equivalent.*

b. *Simple properties of the stereographic projection.* In analogy with the two theorems of § 2, **b**, on the inversion we have here two theorems which justify the application of stereographic projection in circle geometry and related topics:

Theorem B. *The stereographic projection carries circles of the plane into circles on the sphere and conversely; in particular real circles (including straight lines) into real circles on the sphere, imaginary circles into imaginary circles.*

Theorem C. *The stereographic projection is a conformal mapping.*

Proof of Theorem B. Every straight line in the plane is contained in a certain plane through the point **S**; this plane cuts the sphere in a certain circle through **S**, the stereographic image of the line. Conversely every circle through **S** is projected into a straight line in the plane.

For the general case we prove the theorem analytically. Let the circle $\mathfrak{C}$ be given by its equation (1.2). The corresponding curve on the sphere will be obtained by substituting in (1.2)

$$z = \frac{\xi + i\eta}{1+\zeta} \quad \text{with the condition} \quad \xi^2+\eta^2+\zeta^2 = 1.$$

Thus

$$\mathfrak{C}(z, \bar{z}) = A\frac{\xi^2+\eta^2}{(1+\zeta)^2}+B\frac{\xi+i\eta}{1+\zeta}+C\frac{\xi-i\eta}{1+\zeta}+D = 0.$$

Only in the case of a straight line could $1+\zeta$ be equal to zero; this case has been dealt with. Since $\xi^2+\eta^2 = (1+\zeta)(1-\zeta)$ we have

$$(1+\zeta)\mathfrak{C}(z, \bar{z}) = A(1-\zeta)+B(\xi+i\eta)+C(\xi-i\eta)+D(1+\zeta) = 0$$

or

(3.4) $$a\xi+b\eta+c\zeta+d = 0$$

where

(3.41) $$a = B+C, \qquad b = i(B-C), \qquad c = D-A, \qquad d = D+A.$$

These coefficients are all real; therefore (3.4) is the equation of a plane in space.

Does this plane meet the sphere? Its distance from the origin is given by

$$p = \frac{|d|}{\sqrt{(a^2+b^2+c^2)}}.$$

Thus if

$$\left.\begin{array}{l} p^2 < 1 \\ p^2 = 1 \\ p^2 > 1 \end{array}\right\} \text{ the plane (3.4) } \left\{\begin{array}{l} \text{intersects the sphere in a real circle} \\ \text{touches the sphere at a single point} \\ \text{lies entirely outside the sphere.} \end{array}\right.$$

Hence the answer to the question depends on the sign of p^2-1 which is also the sign of

$$\begin{aligned} d^2-a^2-b^2-c^2 &= (D+A)^2-(B+C)^2+(B-C)^2-(D-A)^2 \\ &= 4\Delta = 4|\mathfrak{C}|. \end{aligned}$$

This shows that to every real circle ($\Delta < 0$) corresponds a real circle on the sphere; to every point circle ($\Delta = 0$) a point circle, that is, a single point on the sphere; to every imaginary circle ($\Delta > 0$) a plane which does not meet the sphere ($p^2 > 1$). Any such plane defines an imaginary circle on the sphere.

The converse is immediately clear because the equations (3.41) can be solved for A, B, C, D if a, b, c, d are given:

(3.42) $$A = \tfrac{1}{2}(d-c), \qquad B = \tfrac{1}{2}(a-ib), \qquad C = \tfrac{1}{2}(a+ib), \qquad D = \tfrac{1}{2}(d+c).$$

Remark. The great circles on the sphere are those whose plane passes through the centre of the sphere which is the origin of the coordinate system. They are characterized by the condition $d = 0$. For the corresponding circles $\mathfrak{C}$ in the plane therefore

(3.43) $$D+A = 0.$$

Through every point in the plane a unique circle of this kind exists which touches a given curve at this point.

Proof of Theorem C. The angle Ω between two curves on the sphere, meeting at a point **P**, is defined as the angle between the planes of the two great circles through **P** whose tangents at **P** coincide with the tangents of the two curves at **P**. Let $a_j\xi+b_j\eta+c_j\zeta = 0$ ($j = 1, 2$) be the two planes; then

$$\cos\Omega = \frac{a_1a_2+b_1b_2+c_1c_2}{\sqrt{(a_1^2+b_1^2+c_1^2)}\sqrt{(a_2^2+b_2^2+c_2^2)}}.$$

By (3.41) and (1.51)

$$a_j^2+b_j^2+c_j^2 = -4\Delta_j, \qquad a_1a_2+b_1b_2+c_1c_2 = -4\Delta_{12}.$$

Hence with regard to (1.61): $\cos\Omega = \cos\omega$ where ω is the angle between the stereographic images in the plane of the two great circles. Thus stereographic projection is isogonal.

If the circles on the sphere are viewed from outside, then the sense of rotation is clearly preserved. The projection is therefore a conformal mapping.

Remark. Stereographic projection of orthogonal circles in the z-plane will produce orthogonal circles on the sphere and conversely. The definition of the inversion can thus be transferred to the sphere. Two points **P** and **P*** on the sphere are mutually inverse (or symmetric) with respect to a circle Γ on the sphere, if the circles on the sphere through **P** and **P*** are orthogonal to Γ. Stereographic images of **P** and **P*** in the plane are then two points z and z^* which are mutually inverse with respect to the image $\mathfrak{C}$ of Γ in the plane. Inversion with respect to a great circle on the sphere coincides with ordinary symmetry with respect to the plane of the great circle.

c. *Stereographic projection and polarity.* Let $\mathbf{C} = (\xi_0, \eta_0, \zeta_0)$ be an arbitrary point in space. The plane

$$\xi_0\xi+\eta_0\eta+\zeta_0\zeta = 1 \tag{3.5}$$

is called the *polar plane* γ of **C** with respect to the sphere (3.1) and **C** is called the pole of the *plane* γ (cf. § 2, ex. 6). If **C** is a point on the sphere, then its polar plane coincides with the tangent plane to the sphere at **C**.

Between pole and polar we have the following

Law of Reciprocity. (i) The polar plane of a point **Q** in the plane γ passes through the pole **C** of γ.

(ii) The pole of a plane passing through the point **C** lies in the polar plane γ of **C**.

(i) For any point $\mathbf{Q} = (\xi, \eta, \zeta)$ in γ the polar plane is given by

$$\xi\tilde{\xi}+\eta\tilde{\eta}+\zeta\tilde{\zeta} = 1 \tag{3.51}$$

in running coordinates $\tilde{\xi}, \tilde{\eta}, \tilde{\zeta}$. With respect to (3.5) this equation is satisfied by $\tilde{\xi} = \xi_0, \tilde{\eta} = \eta_0, \tilde{\zeta} = \zeta_0$. Hence **C** lies in the plane (3.51).

(ii) If a plane $a\xi+b\eta+c\zeta = 1$ passes through the point **C**, that is, $a\xi_0+b\eta_0+c\zeta_0 = 1$, then its pole (a, b, c) is a point of the plane (3.5).

If **C** is situated outside the sphere, then its polar plane intersects the sphere in a real circle Γ. By (i) the tangent plane to the sphere at any point **Q** of Γ passes through **C**. Therefore *the pole* **C** *is the vertex of the tangent cone* of the circle Γ.

If **C** lies inside the sphere, then every plane through **C** meets the sphere in a real circle and we obtain the pole of this plane as vertex of the tangent cone. By (i) all these poles lie in the polar plane γ of **C**; three of them are sufficient to determine γ. Similarly by (ii) we may determine the pole **C** of a plane γ that does not meet the sphere.

If the plane γ passes through the origin, viz.

$$a\xi+b\eta+c\zeta = 0, \tag{3.52}$$

then its pole is a point at infinity (in the sense of affine geometry) defined by the normal direction (a, b, c). The polar plane of the centre **O** of the sphere is the plane at infinity of the affine space.

Theorem D. *Let* Γ *be a circle on the sphere* (3.1), γ *its plane, and* **P** *a point on the sphere, but not on* Γ. *Then the inverse* **P*** *of* **P** *with respect to* Γ *is obtained as the second intersection of the straight line through* **C** *and* **P** *with the sphere* (*cf. Figure* 7).

Geometrical Proof. Through every point of Γ draw the great circle orthogonal to Γ and the tangent to this great circle. All these tangents meet at the pole **C** of the plane γ of Γ. Also draw the circles $\tilde{\Gamma}$ orthogonal to Γ and passing through **P** and therefore through **P***. Their tangents at the points of Γ coincide with those of the corresponding great circles. In particular the plane $\tilde{\gamma}$ of a circle $\tilde{\Gamma}$ contains this tangent and therefore **C**. It also contains **P** and **P***. Thus the intersection of the planes $\tilde{\gamma}$, a certain straight line, contains **C**, **P**, and **P***.

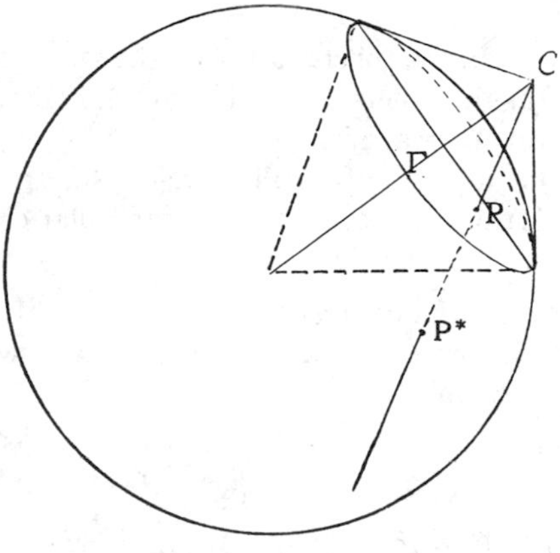

FIG. 7

Remark. This construction of **P*** for a given **P** is valid also if the fundamental circle Γ is imaginary. Its (real) plane γ has then no point in common with the sphere so that the pole **C** lies inside the sphere. An analytical proof will be indicated in ex. 4.

Theorem E. *The circle* $\mathfrak{C}_1$ *is orthogonal to the circle* $\mathfrak{C}_2$ *if and only if the plane* γ_1 *of the corresponding spherical circle* Γ_1 *has its pole* $\mathbf{C}_1$ *in the plane* γ_2 *of* Γ_2.

Proof. (i) If $\mathfrak{C}_1 \perp \mathfrak{C}_2$, then $\Gamma_1 \perp \Gamma_2$. The pole $\mathbf{C}_1$ of γ_1 is obtained as the vertex of the tangent cone of Γ_1. Two of the generating lines of this cone are tangents to Γ_2 and therefore contained in the plane γ_2; hence $\mathbf{C}_1$ lies in γ_2.

(ii) If γ_1 has its pole $\mathbf{C}_1$ within the plane γ_2 of Γ_2, the tangent cone with its vertex at $\mathbf{C}_1$ intersects the plane γ_2 in two tangents to Γ_2, viz. at the two points where it is cut by Γ_1. Let **P** be any point of Γ_2. Then the line $\mathbf{C}_1\mathbf{P}$ meets the sphere a second time at the point **P*** and since this is again a point of Γ_2, by Theorem D the inverse of **P** with respect to Γ_1, we conclude that Γ_1 and Γ_2 must be orthogonal (cf. § 2, ex. 1).

EXAMPLES

1. Let z_1 be the stereographic image of the point $\mathbf{P}_1(\xi_1, \eta_1, \zeta_1)$ on the sphere (3.1). Denote by $\mathbf{P}_2$ the point diametrically opposite to $\mathbf{P}_1$ on the sphere, by z_2 its stereographic image in the plane. Show that z_1 and z_2 are mutually inverse with respect to the imaginary unit circle $(0, i)$. Referring to the position of $\mathbf{P}_1$ and $\mathbf{P}_2$ on the sphere, we call z_1 and z_2 a pair of *antipodal* points.

2. In the z-plane find the stereographic images of the circles which are obtained by intersecting the sphere with a family of parallel planes, for instance, the family of planes perpendicular to the line $\mathbf{P}_1\mathbf{P}_2$ (cf. ex. 1). Show that these circles form a hyperbolic pencil with the 'centres' (point circles) z_1, z_2 (cf. § 2, ex. 4).

3. On the sphere (3.1) find the stereographic image of
(a) the points $\bar{z}$, $-z$, $1/\bar{z}$, $-1/\bar{z}$, if $\mathbf{P}(\xi, \eta, \zeta)$ is the image of z;
(b) the real axis;
(c) a pencil of parallel straight lines;
(d) the segment z_1, z_2; in particular choose $z_1 = 1$, $z_2 = i$;
(e) the triangle z_1, z_2, z_3.

4. *Analytical proof of Theorem D.* Let (3.4) be the equation of the plane γ of the circle Γ. Then the pole $\mathbf{C}$ of γ with respect to the sphere (3.1) has the coordinates

$$-\frac{a}{d}, \quad -\frac{b}{d}, \quad -\frac{c}{d}$$

and if $\mathbf{P} = (\xi, \eta, \zeta)$, then the line through $\mathbf{C}$ and $\mathbf{P}$ is given by $\mathbf{Q} = (1-\lambda)\mathbf{P}+\lambda\mathbf{C}$. If this point has the coordinates ξ_1, η_1, ζ_1 and if it is a point on the sphere (different from $\mathbf{P}$), then

$$\lambda = 2d\frac{a\xi+b\eta+c\zeta+d}{a^2+b^2+c^2+d^2+2d(a\xi+b\eta+c\zeta)}.$$

Now let $z = (\xi+i\eta)/(1+\zeta)$ be the stereographic projection of $\mathbf{P}$ into the z-plane. Further, if $\mathfrak{C}$ is the image of Γ, then by (3.41)

$$\text{(3.6)} \quad \begin{cases} \lambda = \dfrac{(D+A)\mathfrak{C}(z, \bar{z})}{(D+A)\mathfrak{C}(z, \bar{z})-\Delta(1+z\bar{z})} \\[2ex] 1-\lambda = \dfrac{-\Delta(1+z\bar{z})}{(D+A)\mathfrak{C}(z, \bar{z})-\Delta(1+z\bar{z})}. \end{cases}$$

As stereographic projection of $\mathbf{Q}$ one finds

$$z_1 = \frac{\xi_1+i\eta_1}{1+\zeta_1} = \frac{(1-\lambda)(\xi+i\eta)-(\lambda/d)(a+ib)}{1+(1-\lambda)\zeta-(\lambda/d)c}$$

which by means of (3.41) and (3.42) becomes

$$z_1 = \frac{2(1-\lambda)z-2\lambda C(1+z\bar{z})/(D+A)}{[1-\lambda(D-A)/(D+A)](1+z\bar{z})+(1-\lambda)(1-z\bar{z})}.$$

From the values given by (3.6) it follows that

$$z_1 = -\frac{\Delta z + C\mathfrak{C}(z, \bar{z})}{-\Delta + A\mathfrak{C}(z, \bar{z})} = -\frac{C\bar{z}+D}{A\bar{z}+B} = z^*$$

which by (2.21) is indeed the inverse of z with respect to $\mathfrak{C}$. Hence $\mathbf{P}^*$ is the inverse of $\mathbf{P}$ with respect to Γ.

5. In terms of ex. 4, show that the line joining the pole $\mathbf{C}$ to the centre of projection (south pole) $\mathbf{S}$ on the sphere passes through the centre of the circle $\mathfrak{C}$ in the z-plane. This is the Theorem of Chasles. [Hint: Consider the line $(1-\lambda)\mathbf{S}+\lambda\mathbf{C}$].

6. *Polar lines.* If the point $\mathbf{C} = (\xi_0, \eta_0, \zeta_0)$ moves along a certain line $\mathfrak{l}$, then the polar plane γ of $\mathbf{C}$ with respect to the sphere (3.1) turns around a certain line $\mathfrak{l}^*$. This line is called the polar line of $\mathfrak{l}$ with respect to the sphere (3.1). By the reciprocity law it is seen that $\mathfrak{l}$ is the polar line of $\mathfrak{l}^*$. The two lines $\mathfrak{l}$ and $\mathfrak{l}^*$ are perpendicular.

Proof. (i) Let $\mathbf{P}_1 = (\xi_1, \eta_1, \zeta_1)$, $\mathbf{P}_2 = (\xi_2, \eta_2, \zeta_2)$ be two different points on $\mathfrak{l}$. The polar plane of any point $\mathbf{C}$ on $\mathfrak{l}$ is given by the equation (3.5) if we replace ξ_0 by $(1-\lambda)\xi_1+\lambda\xi_2$, etc. Thus all these polar planes contain the intersection of the two planes $\xi_1\xi+\ldots = 1$, $\xi_2\xi+\ldots = 1$ which is the line $\mathfrak{l}^*$.

(ii) If $\mathfrak{l}$ meets the sphere in two points $\mathbf{P}_1$, $\mathbf{P}_2$, the intersection of the tangent planes at $\mathbf{P}_1$ and $\mathbf{P}_2$ is the line $\mathfrak{l}^*$ which is evidently skew and perpendicular to $\mathfrak{l}$. If $\mathfrak{l}$ does not meet the sphere, then there are two planes through $\mathfrak{l}$ that touch the sphere at the two points $\mathbf{P}_1^*$, $\mathbf{P}_2^*$, and $\mathfrak{l}^*$ is the line $\mathbf{P}_1^*\mathbf{P}_2^*$. Finally, suppose that $\mathfrak{l}$ touches the sphere at $\mathbf{P}_1$. Instead of $\mathbf{P}_2$ choose the point at infinity on $\mathfrak{l}$. Its polar plane passes through the centre $\mathbf{O}$ of the sphere and is perpendicular to $\mathfrak{l}$. The polar plane of $\mathbf{P}_1$ is the tangent plane at $\mathbf{P}_1$ and $\mathfrak{l}^*$ is the intersection of these two planes. Hence $\mathfrak{l}^*$ passes through $\mathbf{P}_1$ and is perpendicular to $\mathfrak{l}$.

7. Write down the hermitian matrix $\mathfrak{C}_\infty$ that represents the point circle at the point ∞ in the completed z-plane.

§ 4. Pencils and bundles of circles

a. *Pencils of circles.* In the discussion of the relations between two circles in the plane and later in the definition of the inversion the importance of pencils of circles became apparent. Stereographic projection enables us to transfer to the sphere the geometrical study of pencils. Interpretation of the results in plane geometry is then almost obvious. Let

$$\gamma = (a, b, c, d)$$

be the plane associated with the circle $\mathfrak{C} = \begin{pmatrix} A & B \\ C & D \end{pmatrix}$ according to (3.41) or (3.42). There is thus a one–one correspondence between all circles $\mathfrak{C}$ (including straight lines, imaginary circles, and point circles) in the completed plane and the planes γ of the affine space[3] (including the plane (0, 0, 0, 1) at infinity to which corresponds the circle $\mathfrak{E} = \begin{pmatrix} 1 & 0 \\ 0 & 1 \end{pmatrix}$, that is, the imaginary unit circle).

Now let $\gamma_j = (a_j, b_j, c_j, d_j)(j = 1, 2)$ be two distinct planes and $\mathfrak{C}_j = \begin{pmatrix} A_j & B_j \\ C_j & D_j \end{pmatrix}$ the associated circles. By

$$\mathfrak{l} = \langle \gamma_1, \gamma_2 \rangle$$

we denote the line of intersection of γ_1 and γ_2. All planes through $\mathfrak{l}$ form a pencil of planes.

$$\gamma = \lambda_1 \gamma_1 + \lambda_2 \gamma_2 = (\lambda_1 a_1 + \lambda_2 a_2, \ldots, \lambda_1 d_1 + \lambda_2 d_2).$$

The line $\mathfrak{l}$ is called the axis of the pencil (γ). By intersection of the planes with the sphere a pencil of circles on the sphere is obtained, to which corresponds the pencil ($\mathfrak{C}$) consisting of the circles $\mathfrak{C} = \lambda_1 \mathfrak{C}_1 + \lambda_2 \mathfrak{C}_2$ in the completed plane.

Three different cases are to be distinguished:

(i) Suppose that the axis $\mathfrak{l}$ intersects the sphere in two different points $\mathbf{P}_1$, $\mathbf{P}_2$. Then the pencil ($\mathfrak{C}$) is *elliptic* and consists of all circles passing through the stereographic images z_1, z_2 of $\mathbf{P}_1$, $\mathbf{P}_2$. If $\mathbf{P}_1 = \mathbf{S}$, then the pencil ($\mathfrak{C}$) consists of straight lines only, viz. all lines through the point z_0, where the axis $\mathfrak{l}$ intersects the z-plane.

(ii) If $\mathfrak{l}$ touches the sphere at $\mathbf{P}_1$, the pencil ($\mathfrak{C}$) will be *parabolic*. The pencils on the sphere touch each other and the line $\mathfrak{l}$. Between the circles of the pencil ($\mathfrak{C}$) there is, in general, one straight line, corresponding to the plane through the point $\mathbf{S}$. This line is defined by the two points z_0 and z_1, the stereographic image of $\mathbf{P}_1$. If $\mathbf{P}_1 = \mathbf{S}$, then all the circles $\mathfrak{C}$ of the pencil are parallel straight lines.

(iii) If $\mathfrak{l}$ does not meet the sphere, the pencil ($\mathfrak{C}$) will be *hyperbolic*. To those two planes of the pencil (γ) which touch the sphere, correspond the two point circles of the pencil ($\mathfrak{C}$). However γ_1, γ_2 may be chosen, the pencil ($\mathfrak{C}$) will always contain real circles (cf. § 1, ex. 7). The plane through $\mathbf{S}$ will produce in the z-plane the only straight line of the pencil ($\mathfrak{C}$) unless the plane touches the sphere at $\mathbf{S}$. In this case, if another plane touches the sphere at $\mathbf{Q}$, then ($\mathfrak{C}$) consists of concentric circles about the stereographic image z_2 of $\mathbf{Q}$. (This follows from § 2, ex. 4 and also from the following Theorem A.)

Theorem A. *For every hyperbolic (parabolic, elliptic) pencil* ($\mathfrak{C}$) *there is a unique elliptic (parabolic, hyperbolic) pencil* ($\tilde{\mathfrak{C}}$) *such that every circle of* ($\tilde{\mathfrak{C}}$) *is orthogonal to every circle of* ($\mathfrak{C}$).

[3] The plane-coordinates a, b, c, d as well as the circle-coordinates A, B, C, D are homogeneous coordinates; thus they cannot vanish all four at the same time and a plane or circle (cf. § 1, a) defines its coordinates only up to proportionality so that one of the non-zero coordinates may always be chosen to be equal to 1.

Proof. Let $(\mathfrak{C})$ be hyperbolic, the axis therefore entirely outside the sphere. Let γ_1, γ_2 be the two planes through $\mathfrak{l}$, which touch the sphere at the two points $\mathbf{Q}_1$, $\mathbf{Q}_2$. The line $\tilde{\mathfrak{l}}$ through $\mathbf{Q}_1$ and $\mathbf{Q}_2$ will be the polar to $\mathfrak{l}$ (cf. § 3, ex. 6). The pencil of planes $\tilde{\gamma}$ with the axis $\tilde{\mathfrak{l}}$ will produce an elliptic pencil $(\tilde{\mathfrak{C}})$ in the z-plane. Every plane $\tilde{\gamma}$ will have its pole on the line $\mathfrak{l}$, and each plane γ will have its pole within every $\tilde{\gamma}$ and therefore on $\tilde{\mathfrak{l}}$. Hence by § 3, Theorem E, the circles $\mathfrak{C}$ and $\tilde{\mathfrak{C}}$ are orthogonal. The same argument applies if we start with a parabolic or an elliptic pencil $(\mathfrak{C})$.

Corollary. *The two points through which pass all the circles of an elliptic pencil are the point circles of the associated orthogonal hyperbolic pencil.*

The configuration of two orthogonal pencils of circles is of fundamental importance throughout the present exposition (cf. § 8, **c**).

b. *Bundles of circles.* All planes passing through one point $\mathbf{P}$ in space form a bundle of planes. If γ_1, γ_2, γ_3 are three different planes of the bundle with the centre $\mathbf{P}$ not all three in one and the same pencil, then

$$\gamma = \lambda_1\gamma_1+\lambda_2\gamma_2+\lambda_3\gamma_3 \qquad (\lambda_j \text{ real, not all zero})$$

represents the general plane of the bundle. It contains every pencil generated by two of its planes. If $\mathbf{P}$ is an infinite point of the space, then the bundle consists of all pencils whose axes $\mathfrak{l}$ pass through $\mathbf{P}$, that is, are parallel to a certain direction in space.

By its intersections the bundle of planes defines a bundle of circles on the sphere which by stereographic projection is turned into a bundle of circles in the z-plane:

$$\mathfrak{C} = \lambda_1\mathfrak{C}_1+\lambda_2\mathfrak{C}_2+\lambda_3\mathfrak{C}_3.$$

Theorem B. *For every bundle of circles $\mathfrak{C}$ there is a circle $\mathfrak{P}$ such that every circle $\mathfrak{C}$ is orthogonal to $\mathfrak{P}$.*

Proof. Denote by π the polar plane of the centre $\mathbf{P}$ of the bundle of planes γ. The plane π then contains the poles with respect to the sphere of all these planes γ. The circles Π and Γ in which these planes intersect the sphere, are therefore orthogonal (§ 3, Theorem E) and so are the corresponding circles $\mathfrak{P}$ and $\mathfrak{C}$ in the z-plane.

The geometrical nature of a bundle depends on the position of its centre $\mathbf{P}$ with respect to the sphere. Thus, as in the case of the pencils, the following three cases occur:

(i) *Elliptic bundle.* The point $\mathbf{P}$ lies inside the sphere; thus all circles of the bundle are real (cf. § 1, ex. 11) and all its pencils are elliptic. Since the polar plane of $\mathbf{P}$ lies entirely outside the sphere, the circle $\mathfrak{P}$ of Theorem B is imaginary. If $\mathbf{P}$ coincides with the centre of the sphere, then the bundle of circles Γ on the sphere coincides with the system of all great circles. The corresponding bundle in the z-plane is defined by the equation (3.43).

(ii) *Parabolic bundle.* The point **P** lies on the sphere and all circles of the bundle are real; they are the circles passing through **P**. The axes $\mathfrak{l}$ of the pencils contained in the bundle either touch the sphere or intersect it at **P**; the corresponding pencils of circles are therefore either parabolic or elliptic. If **P** = **S** we obtain as bundle the system of all straight lines in the z-plane. The circle $\mathfrak{P}$ of Theorem B is then the point circle at infinity.

(iii) *Hyperbolic bundle.* The point **P** lies outside the sphere. The bundle contains real as well as imaginary circles and an infinity of point circles. These correspond to the planes γ that touch the sphere; they are therefore situated on the circle $\mathfrak{P}$ which is obviously a real circle. A hyperbolic bundle contains elliptic, parabolic, and hyperbolic pencils. If **P** is shifted into the infinite point of the ζ-axis, then the bundle consists of all circles which are orthogonal to the equator, that is, the unit circle about the origin in the z-plane.

For each of the three types of bundles we have given a representative; later we shall see (III, § 12, **c**) that certain evident properties of these representatives are in fact characteristic for all bundles of their own types. For this purpose we shall have to use a more general class of transformations of the z-plane onto itself by which, as in the case of the inversions, circles are carried into circles.

EXAMPLES

1. Show that by inversion every elliptic (parabolic, hyperbolic) pencil or bundle is transformed into an elliptic (parabolic, hyperbolic) pencil or bundle.

2. Show that by a suitable inversion two given real circles in the z-plane can be taken either into a pair of straight lines or into a pair of concentric circles.

Hence every pencil of circles can be transformed by a suitable inversion into one of the following three 'normal forms':

(i) Elliptic: the pencil of all straight lines through a point z_0.

(ii) Parabolic: a pencil of parallel straight lines.

(iii) Hyperbolic: the pencil of all concentric circles about a point z_0.

In what way is the point z_0 in the cases (i) and (iii), and the direction of the parallel lines in the case (ii), restricted by the given pencil?

3. All circles $\mathfrak{C}$ orthogonal to two given circles $\mathfrak{C}_1$, $\mathfrak{C}_2$ form a pencil. Carry through the classification of pencils on the basis of this definition.

4. A system of circles is a bundle if there is a point γ_0 having the same power with respect to each of these circles (cf. § 2, ex. 7). Carry through the classification of bundles on the basis of this definition.

5. Every bundle of circles $\mathfrak{C}$ may be characterized by one linear equation in the elements A, B, C, D of the hermitian matrix $\mathfrak{C}$, with real coefficients (cf., for example, (3.43)). From this statement it follows that the system of all real circles with the same radius is not a bundle.

6. For every pencil or bundle, a real circle (line) $\mathfrak{C}_0$ can be found such that the pencil or bundle is transformed into itself by inversion with respect to $\mathfrak{C}_0$.

7. For every bundle of circles $\mathfrak{C}$ a real circle $\mathfrak{C}_1$ can be found such that inversion with respect to $\mathfrak{C}_i$ carries the bundle into one of the three 'normal forms':

(i) Elliptic: (3.43) $A+D = 0$.

(ii) Parabolic: all straight lines: $A = 0$.

(iii) Hyperbolic: all circles orthogonal to the unit circle: $A-D = 0$.

8. For two given circles $\mathfrak{C}_1$ and $\mathfrak{C}_2$ of the elliptic pencil through γ_1, γ_2 find by construction with ruler and compass the circles $\mathfrak{C}$ and $\tilde{\mathfrak{C}}$ of the pencil which bisect the angles between $\mathfrak{C}_1$ and $\mathfrak{C}_2$.

Algebraically: Let $\mathfrak{C} = \lambda_1\mathfrak{C}_1+\lambda_2\mathfrak{C}_2$; let ω_j be the angle between $\mathfrak{C}$ and $\mathfrak{C}_j$ $(j = 1, 2)$. Using the notations of § 1 and putting

$$\Delta_{0j} = A_j(\lambda_1 D_1+\lambda_2 D_2)+D_j(\lambda_1 A_1+\lambda_2 A_2)-B_j(\lambda_1 C_1+\lambda_2 C_2)-C_j(\lambda_1 B_1+\lambda_2 B_2)$$

we may write the condition $\cos\omega_1 = \cos\omega_2$ in the form

$$\frac{\Delta_{01}}{\sqrt{-\Delta_1}} = \pm\frac{\Delta_{02}}{\sqrt{-\Delta_2}}$$

whence

$$\lambda = \frac{\lambda_2}{\lambda_1} = \pm\frac{\sqrt{-\Delta_1}}{\sqrt{-\Delta_2}}.$$

By means of § 1, ex. 12, distinction may be made between the two circles $\mathfrak{C}_1+\lambda\mathfrak{C}_2$ bisecting the two angles between $\mathfrak{C}_1$ and $\mathfrak{C}_2$. From (2.72) (§ 2, ex. 10) it is evident that $\mathfrak{C}_1$ and $\mathfrak{C}_2$ are mutually symmetric with respect to each of the two circles $\mathfrak{C} = \mathfrak{C}_1+\lambda\mathfrak{C}_2$.

§ 5. The cross ratio

a. *The simple ratio.* In order to avoid exceptions in geometrical theorems it was convenient to adopt the symbol ∞ as an additional point in the z-plane. By stereographic projection to the sphere, ∞, as well as every 'finite' point z, had a certain image point on the sphere; thus it was justified that ∞ should not be considered as essentially distinct from any other point in the completed plane. This, of course, does not mean that ∞ should be taken as a new element of the field of all complex numbers. The algebraic laws in this field, however, can be extended so as to cover certain operations involving the symbol ∞ without producing a contradiction to ordinary algebra.

Let c be a complex number or ∞; then we define

$$(5.1)\qquad \begin{cases} \text{For } c \neq \infty: & c\pm\infty = \infty, \quad \dfrac{c}{\infty} = 0. \\[2ex] \text{For } c \neq 0: & c\cdot\infty = \infty, \quad \dfrac{c}{0} = \infty. \end{cases}$$

The expressions $\infty \pm \infty$, ∞/∞, $0\cdot\infty$, $0/0$ are not defined.

We may define the symbol ∞ as the limit of any sequence of complex numbers $c_1, c_2, \ldots$, which possesses no finite point of accumulation (which is the case if and only if the absolute values $|c_n|$ tend to infinity as $n \to \infty$, so that every circle about 0 contains only a finite number of the c_n). If we accept any such sequence as convergent, we can apply to it the ordinary laws of the algebra of limits of convergent sequences, (for example, $\lim c_n + \lim c_n' = \lim(c_n + c_n')$, etc.). In this way the rules (5.1) can readily be proved.

Now let z_1, z_2, z_3 be three complex numbers, not all three equal. Then the simple ratio of $z_1; z_2, z_3$ will be defined by

$$(5.2) \qquad (z_1; z_2, z_3) = \begin{cases} \dfrac{z_1 - z_2}{z_1 - z_3} & \text{if } z_3 \neq z_1 \\ \infty & \text{if } z_3 = z_1. \end{cases}$$

We complete this definition by defining

$$(5.21) \qquad (\infty; z_2, z_3) = 1, \qquad (z_1; \infty, z_3) = \infty, \qquad (z_1; z_2, \infty) = 0$$

and

$$(5.22) \qquad (\infty; \infty, z_3) = 0, \qquad (\infty; z_2, \infty) = \infty, \qquad (z_1; \infty, \infty) = 1.$$

Then, if $z_1; z_2, z_3$ is any triplet of points of the completed complex plane, not all three equal, the following relations hold:

$$(5.3) \qquad \begin{cases} (z_1; z_2, z_3) = r, & (z_2; z_3, z_1) = \dfrac{r-1}{r}, & (z_3; z_1, z_2) = \dfrac{1}{1-r} \\ (z_1; z_3, z_2) = \dfrac{1}{r}, & (z_2; z_1, z_3) = \dfrac{r}{r-1}, & (z_3; z_2, z_1) = 1 - r. \end{cases}$$

It may be pointed out that the definitions (5.21) and (5.22) are deliberately chosen so that the permutation rules (5.3) remain valid also if one or two of the z_j are ∞. Since by (5.21)

$$\lim_{z \to \infty} (\infty; z, z_3) = 1 \neq (\infty; \infty, z_3)$$

it follows that $(\infty; z, z_3)$ is discontinuous at $z = \infty$. It turns out to be impossible to have the permutation rules (5.3) intact and at the same time the simple ratio continuous at infinity. Other values may be chosen instead of those given in (5.22).

b. *The double ratio or cross ratio.* Let z_1, z_2, z_3, z_4 be four points of the completed plane no three of which are equal; then the four simple ratios

$$(z_1; z_3, z_4), \qquad (z_2; z_3, z_4), \qquad (z_3; z_1, z_2), \qquad (z_4; z_1, z_2)$$

exist. Moreover we assume that neither the first two, nor the last two of these ratios are simultaneously equal to 0 or to ∞. We define the *cross ratio* (also called double ratio or anharmonic ratio) of the two pairs $z_1, z_2; z_3, z_4$ by

$$(5.4) \qquad (z_1, z_2; z_3, z_4) = \frac{(z_1; z_3, z_4)}{(z_2; z_3, z_4)} = \frac{z_1-z_3}{z_1-z_4} \div \frac{z_2-z_3}{z_2-z_4}.$$

The four permutations of the four-group applied to the four points z_1, z_2, z_3, z_4 do not alter the value of the cross ratio:

$$(5.41)\ (z_1, z_2; z_3, z_4) = (z_2, z_1; z_4, z_3) = (z_3, z_4; z_1, z_2) = (z_4, z_3; z_2, z_1) = \lambda.$$

Hence there are at most six values which the cross ratio can assume if the points z_1, z_2, z_3, z_4 are subjected to other permutations. As in the case of the simple ratio (cf. (5.3)) the five other values are

$$(5.42) \qquad (z_2, z_3; z_1, z_4) = \frac{\lambda-1}{\lambda}, \qquad (z_3, z_1; z_2, z_4) = \frac{1}{1-\lambda},$$

$$(z_2, z_1; z_3, z_4) = \frac{1}{\lambda}, \qquad (z_3, z_2; z_1, z_4) = \frac{\lambda}{\lambda-1}, \qquad (z_1, z_3; z_2, z_4) = 1-\lambda.$$

These will be obtained immediately from (5.3).

First it will be shown that of the restrictions assumed for the definition (5.4) the one that the ratios be not both equal to zero or infinite can be omitted. In fact let

$$(z_1; z_3, z_4) = (z_2; z_3, z_4) = \infty.$$

Because no three points in the quadruplet may be equal to each other, this cannot be due to $z_1 = z_4$ and $z_2 = z_4$; hence it follows that $z_3 = \infty$ and if z_1, z_2, z_4 are finite, then

$$(5.43) \qquad (z_1, z_2; \infty, z_4) = (\infty, z_4; z_1, z_2) = \frac{(\infty; z_1, z_2)}{(z_4; z_1, z_2)} = (z_4; z_2, z_1)$$

by (5.21). If z_1 or z_2 is infinite, the formula (5.43) gives the value 0 or ∞ respectively. These are also found to be the limits of $(z, z_2; \infty, z_4)$ and $(z_1, z; \infty, z_4)$ as $z \to \infty$.

Similarly if $(z_1; z_3, z_4) = (z_2; z_3, z_4) = 0$ one concludes that $z_4 = \infty$ and therefore

$$(5.44) \qquad (z_1, z_2; z_3, \infty) = \frac{(z_3; z_1, z_2)}{(\infty; z_1, z_2)} = (z_3; z_1, z_2).$$

This formula is also valid if $z_1 = \infty$ or $z_2 = \infty$, giving the values ∞ or 0 respectively for the cross ratio. Moreover $(z, z_2; z_3, \infty)$ and $(z_1, z; z_3, \infty)$ are continuous in z for $z = \infty$.

Thus the cross ratio is defined for any set of four points of the completed plane, if not more than two of the four points are equal. Further, if three of the four points are kept constant, the cross ratio is a continuous function of the variable fourth point, also at infinity.

With regard to this result the permutation rules (5.42) may be obtained by (5.3) from (5.44) which expresses $(z_1, z_2; z_3, z_4)$ for $z_4 = \infty$ as a simple ratio.

Remark. In the geometrical discussion of §§ 1–4 the complex number notation provided a convenient alternative for the cartesian coordinate method. From now on the complex numbers will have a deeper significance. The notion of cross ratio has its origin in real projective geometry where it is defined for a set of four points on a straight line, for instance on the x-axis. This line (as well as any other line) may be considered as carrier of the field of all real numbers which, for the purposes of real projective geometry has been completed by one point at infinity. For a set of four non-collinear points in the plane of projective geometry the cross ratio is not defined. It is just the association of the points of the plane with the complex numbers and the completion of the plane by the *one* point at infinity, which enables us to extend the definition of the cross ratio to sets of four points in the completed plane. This suggests another interpretation of the completed z-plane, viz. as the complex projective straight line. Indeed, every field (in the sense of algebra), completed by an infinite element, may be taken as a projective straight line and thus as the basis for an abstract projective geometry.[4] In this sense, to be specified later on (cf. § 11, **b**), our geometry of the complex numbers then appears as one-dimensional complex projective geometry.[5]

c. *The cross ratio in circle geometry.*

Theorem A. *Four points z_1, z_2, z_3, z_4 of the completed plane lie on one and the same circle if and only if their cross ratio $(z_1, z_2; z_3, z_4)$ is real.*

Proof. If three of the points, for instance, z_2, z_3, z_4, are not mutually different, then $(z_1, z_2; z_3, z_4)$ is 0, or 1, or ∞ (which as a point on the real axis is counted to be real), and there is a circle through these points.

From now on let z_2, z_3, z_4 be mutually different. If one of them is infinite, for example, $z_4 = \infty$, then by (5.44) $(z_1, z_2; z_3, z_4)$ is really a simple ratio and it is readily seen that three finite points, z_1, z_2, z_3 lie on a straight line if and only if the ratio $(z_3; z_1, z_2)$ is real.

Thus we may assume that z_2, z_3, z_4 are finite. Then $(z, z_2; z_3, z_4)$ is real:

$$(\bar{z}, \bar{z}_2; \bar{z}_3, \bar{z}_4) = (z, z_2; z_3, z_4)$$

if and only if

$$(5.5)\qquad (z-z_3)(\bar{z}-\bar{z}_4)(z_2-z_4)(\bar{z}_2-\bar{z}_3) - (z-z_4)(\bar{z}-\bar{z}_3)(z_2-z_3)(\bar{z}_2-\bar{z}_4) = 0.$$

4 Cf. G. de B. Robinson [1], chapters VI, VII, IX.

5 An extensive treatment of this subject has been given by E. Cartan [2], Part I.

After multiplying this equation by i it becomes an equation of the form (1.2) where, by our assumptions, not all the coefficients vanish. Hence (5.5) represents a circle. Since z_2, z_3, z_4 lie on this circle, (5.5) is the equation of the unique real circle through these three points.

A second, more elementary proof runs as follows.

We consider only the case that z_2, z_3, z_4 are finite and non-collinear. Let

$$z_j - z_k = \rho_{jk} e^{i\phi_{jk}} \qquad \rho_{jk} = |z_j - z_k| \qquad (j = 1, 2;\ k = 3, 4)$$

so that

$$\alpha_1 = \phi_{13} - \phi_{14}, \qquad \alpha_2 = \phi_{23} - \phi_{24}$$

are the angles under which the segment with the endpoints z_3 and z_4 appears from z_1, z_2 respectively. Then

$$(z_1; z_3, z_4) = \rho_1 e^{i\alpha_1}, \qquad (z_2; z_3, z_4) = \rho_2 e^{i\alpha_2}, \qquad \rho_j = \frac{\rho_{j3}}{\rho_{j4}}$$

and

$$(z_1, z_2; z_3, z_4) = \frac{\rho_1}{\rho_2} e^{i(\alpha_1 - \alpha_2)}.$$

This is real if and only if $\alpha_1 - \alpha_2$ is an integral multiple of π, which according to a theorem of elementary geometry is the case if and only if z_1 and z_2 lie on the same circle over the chord z_3, z_4 (cf. Figure 8).

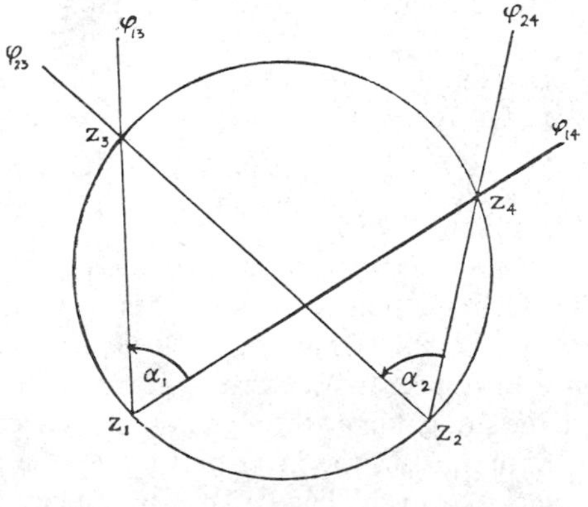

FIG. 8

Another proof is contained in the following ex. 3. Cf. also § 6, **e**.

We conclude with the following application of Theorem A. Let z_2, z_3, z_4 be three mutually different complex numbers. Let λ denote a real parameter. Then

$$(5.6) \qquad (z, z_2; z_3, z_4) = \lambda$$

is a linear equation in z. By solving with respect to z we obtain a parameter representation of the circle through z_2, z_3, z_4 so that each λ determines a point z on this circle $\mathfrak{C}$; conversely every z on $\mathfrak{C}$ determines a value λ. Thus (5.6) is a continuous one–one mapping of $\mathfrak{C}$ onto the completed real axis.

The three points z_2, z_3, z_4 divide the circle $\mathfrak{C}$ into three arcs z_3z_2, z_2z_4, z_4z_3. If

$$\lambda \text{ increases from} \begin{Bmatrix} 0 & \text{to} & 1 \\ 1 & \text{to} & \infty \\ -\infty & \text{to} & 0 \end{Bmatrix}, \; z \text{ moves on} \begin{cases} z_3z_2 \text{ from } z_3 \text{ to } z_2 \\ z_2z_4 \text{ from } z_2 \text{ to } z_4 \\ z_4z_3 \text{ from } z_4 \text{ to } z_3 \end{cases}$$

(note that $\lambda = (\lambda, 1; 0, \infty)$).

In particular $\lambda = -1$ corresponds to a point z_1 on the arc z_4z_3. The points $z_1, z_2; z_3, z_4$ for which

$$(5.61) \qquad (z_1, z_2; z_3, z_4) = -1$$

are said to separate each other harmonically on $\mathfrak{C}$ or to form a harmonic set or a harmonic division of $\mathfrak{C}$. In particular, $-1, 1; 0, \infty$ separate each other harmonically on the real axis.

Theorem B. *Let z_j^* be the inverse point of $z_j (j = 1, 2, 3, 4)$ with respect to a circle $\mathfrak{C}_0$. Then*

$$(z_1^*, z_2^*; z_3^*, z_4^*) = \overline{(z_1, z_2; z_3, z_4)}.$$

Proof. In virtue of (2.31) we may assume that $\mathfrak{C}_0$ is a real circle. Thus by (2.22) (omitting the case of a straight line (cf. ex. 2))

$$z_j^* = \gamma_0 + \frac{\rho_0^2}{\bar{z}_j - \bar{\gamma}_0}, \qquad z_j^* - z_k^* = \rho_0^2 \frac{\bar{z}_k - \bar{z}_j}{(\bar{z}_j - \bar{\gamma}_0)(\bar{z}_k - \bar{\gamma}_0)}$$

$$(z_j^*; z_3^*, z_4^*) = (\bar{\gamma}_0; \bar{z}_4, \bar{z}_3)(\bar{z}_j; \bar{z}_3, \bar{z}_4) \qquad (j = 1,2)$$

and the theorem follows immediately.

Hence one derives a simple geometrical construction of the fourth harmonic z_4 to three mutually different, non-collinear points z_1, z_2, z_3. Let $\mathfrak{C}$ be the circle through these three points. Draw the circle $\mathfrak{C}_0$ about z_3 through z_1 (cf. Figure 9). Inversion with respect to $\mathfrak{C}_0$ carries z_3 into $z_3^* = \infty$, leaves z_1 fixed, and thus maps $\mathfrak{C}$ onto the straight line $\mathfrak{l}$ through z_1 and the second intersection z_1' of $\mathfrak{C}$ and $\mathfrak{C}_0$. Further we find z_2^*, the inverse of z_2, as the intersection of the diameter through z_2 and z_3 with the line $\mathfrak{l}$. Thus by Theorem B one has for z_4^* the condition

$$(z_1, z_2^*; \infty, z_4^*) = -1$$

whence

$$z_4^* = \tfrac{1}{2}(z_1 + z_2^*).$$

This point is readily found on $\mathfrak{l}$. Now join z_3 and z_4^* by a straight line; this line will intersect $\mathfrak{C}$ at z_4.

EXAMPLES

1. The ratio $(z_1;\, z_2, z_3)$ is real if and only if the three points z_1, z_2, z_3 are collinear.

2. If z_1^*, z_2^*, z_3^* are obtained from z_1, z_2, z_3 by inversion with respect to a straight line, then $(z_1^*;\, z_2^*, z_3^*) = \overline{(z_1;\, z_2, z_3)}$.

3. For every (real) circle $\mathfrak{C}$ another circle $\mathfrak{C}_0$ can be found such that inversion with respect to $\mathfrak{C}_0$ carries $\mathfrak{C}$ into the real axis.

Together with Theorem B this yields another proof of Theorem A.

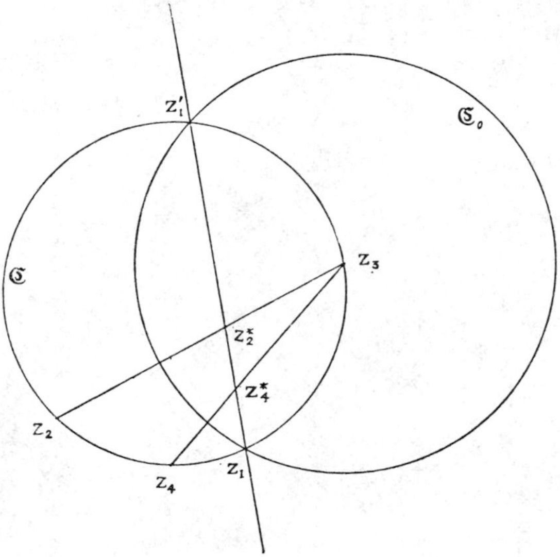

FIG. 9

4. Let

$$z_j = \gamma_0 + \rho_0 e^{i\alpha_j} \qquad (j = 1, 2, 3, 4)$$

be four points on a circle. Show that

$$(z_1, z_2;\, z_3, z_4) = \frac{\sin\frac{1}{2}(\alpha_1-\alpha_3)}{\sin\frac{1}{2}(\alpha_1-\alpha_4)} \div \frac{\sin\frac{1}{2}(\alpha_2-\alpha_3)}{\sin\frac{1}{2}(\alpha_2-\alpha_4)}.$$

5. Describe the possible relative positions of three points z_1, z_2, z_3 for which two of the six values of the ratio $(z_j;\, z_k, z_l)$ are equal to each other.

6. Describe the relative position of four points z_1, z_2, z_3, z_4 for which two of the six values of the cross ratio $(z_j, z_k;\, z_l, z_m)$ are equal. What are the possible values of the cross ratio? Analyse the relation $(z_1, z_2;\, z_3, z_4) = 1$.

7. The construction of the fourth harmonic as described under Theorem B fails if z_1, z_2, z_3 are collinear. Complete it in this case making use of the fact that if z_1 and z_2 are symmetric with respect to a real circle that meets the line through z_1 and z_2 at z_3 and z_4, then $(z_1, z_2; z_3, z_4) = -1$.

8. For three given points z_1, z_2, z_3 find by geometrical construction a fourth point z_4 such that $(z_1, z_2; z_3, z_4)$ has a prescribed real or complex value.

9. Four points z_1, z_2, z_3, z_4 lie on one and the same circle if and only if

$$\begin{vmatrix} z_1\bar{z}_1 & z_1 & \bar{z}_1 & 1 \\ z_2\bar{z}_2 & z_2 & \bar{z}_2 & 1 \\ z_3\bar{z}_3 & z_3 & \bar{z}_3 & 1 \\ z_4\bar{z}_4 & z_4 & \bar{z}_4 & 1 \end{vmatrix} = 0 \qquad \text{[cf. (5.5)].}$$

II

THE MOEBIUS TRANSFORMATION

§ 6. Definition: elementary properties

a. *Definition and notation.* A Moebius transformation is given by

$$Z = \mathfrak{H}(z) = \frac{az+b}{cz+d} \tag{6.1}$$

where z may vary throughout the completed complex plane. The elements of the matrix

$$\mathfrak{H} = \begin{pmatrix} a & b \\ c & d \end{pmatrix}$$

are constant complex numbers. Its determinant

$$|\mathfrak{H}| = \delta = ad-bc$$

is not zero, for if $|\mathfrak{H}| = 0$, then the right-hand side of (6.1) is constant.

The matrix $\mathfrak{H}$ evidently defines the Moebius transformation (6.1) whereas the transformation defines the matrix only up to a constant factor so that all matrices $q\mathfrak{H}$ with any complex $q \neq 0$ define one and the same Moebius transformation.

Like the inversion the Moebius transformation is a one–one correspondence between the z-plane and the Z-plane, where the latter may be considered as a second copy of the z-plane, (cf. § 2, **b**). If both planes (including the coordinate systems and the scales on the axes) coincide we shall say that the function (or transformation) (6.1) represents a mapping of the completed plane onto itself. To different values of z correspond different values of Z, and conversely. The point

$$z_\infty = -\frac{d}{c}$$

is called the pole of the function $\mathfrak{H}(z)$; its image in the Z-plane is the point $Z = \infty$.

Remark 1. A Moebius transformation is also called a *homography* or a *linear transformation* in the variable z, the latter for the following reason. Instead of the variables z; Z we may introduce homogeneous variables z_1, z_2; Z_1, Z_2 respectively so that

$$z = \frac{z_1}{z_2}, \qquad Z = \frac{Z_1}{Z_2}.$$

The relation (6.1) then becomes

$$\text{(6.11)}\qquad \begin{aligned} \frac{1}{q}Z_1 &= az_1+bz_2 \\ \frac{1}{q}Z_2 &= cz_1+dz_2 \end{aligned}$$

where q is an arbitrary complex number different from zero. This is a linear homogeneous transformation in the variables z_1, z_2 with the matrix $\mathfrak{H}$. Making use of the column notation

$$\mathfrak{z} = \begin{pmatrix} z_1 \\ z_2 \end{pmatrix}, \quad \mathfrak{Z} = \begin{pmatrix} Z_1 \\ Z_2 \end{pmatrix}$$

we may write (6.11) in the form

$$\text{(6.12)}\qquad \mathfrak{Z} = q\mathfrak{H}\mathfrak{z} \qquad (q \neq 0).$$

Finally (6.1) is often called a *bilinear transformation* because it is derived from the 'bilinear relation' between the two variables z and Z, viz.

$$\text{(6.13)}\qquad czZ+dZ-az-b = 0,$$

that is, an implicit representation of the function (6.1).

Remark 2. An inversion (2.21) is not a Moebius transformation. It is a special case of a transformation of the form

$$Z = \mathfrak{H}(\bar{z}) \qquad (|\mathfrak{H}| \neq 0),$$

which is called an *anti-homography*. These anti-homographies will be dealt with in § 9.

b. *The group of all Moebius transformations.* Let $\mathfrak{H}_1$, $\mathfrak{H}_2$ be two Moebius transformations:

$$\mathfrak{H}_1 = \begin{pmatrix} a_1 & b_1 \\ c_1 & d_1 \end{pmatrix}, \quad \mathfrak{H}_2 = \begin{pmatrix} a_2 & b_2 \\ c_2 & d_2 \end{pmatrix}.$$

If these transformations are carried out in succession, first $z_1 = \mathfrak{H}_1(z)$, then $Z = \mathfrak{H}_2(z_1)$, we obtain the transformation

$$\text{(6.2)}\qquad Z = \mathfrak{H}_2[\mathfrak{H}_1(z)],$$

which is called the *product* of the two transformations $\mathfrak{H}_1$, $\mathfrak{H}_2$ in this order. It is readily seen to be again a Moebius transformation, viz.

$$Z = \frac{a_2\mathfrak{H}_1(z)+b_2}{c_2\mathfrak{H}_1(z)+d_2} = \frac{(a_2a_1+b_2c_1)z+a_2b_1+b_2d_1}{(c_2a_1+d_2c_1)z+c_2b_1+d_2d_1} = \mathfrak{H}_3(z)$$

whose matrix $\mathfrak{H}_3$ appears as the product of the two matrices $\mathfrak{H}_1$, $\mathfrak{H}_2$:

$$\mathfrak{H}_3 = \mathfrak{H}_2\mathfrak{H}_1 = \begin{pmatrix} a_2a_1+b_2c_1 & a_2b_1+b_2d_1 \\ c_2a_1+d_2c_1 & c_2b_1+d_2d_1 \end{pmatrix}. \tag{6.21}$$

The product function (6.2) is certainly not constant since neither of the functions $\mathfrak{H}_1(z)$, $\mathfrak{H}_2(z)$ is assumed to be constant so that $|\mathfrak{H}_3| = |\mathfrak{H}_2||\mathfrak{H}_1| \neq 0$.

The Moebius transformation (6.1) is invertible; its inverse is again a Moebius transformation

$$z = \mathfrak{H}^{-1}(Z) = \frac{dZ-b}{-cZ+a}. \tag{6.22}$$

Its matrix is the inverse matrix $\mathfrak{H}^{-1}$ of $\mathfrak{H}$, or, omitting the numerical factor $1/\delta$ (as in (6.22)) the matrix $\begin{pmatrix} d & -b \\ -c & a \end{pmatrix}$.

A special Moebius transformation is the identity transformation $Z = z$; its matrix is the unit matrix

$$\mathfrak{E} = \begin{pmatrix} 1 & 0 \\ 0 & 1 \end{pmatrix}.$$

Matrix multiplication is associative; the same is therefore true for the composition of three or more Moebius transformations.[1]

The preceding statements may be combined into

Theorem A. *The system of all Moebius transformations is a group with functional composition as group multiplication.*

By means of stereographic projection every Moebius transformation can be transferred to the sphere. It appears there as a certain transformation of the sphere onto itself. The group of all these spherical transformations is isomorphic to the group of all Moebius transformations. The analytical representation of the spherical transformations turns out to be rather complicated in general (cf. § 11, **c**). An important special case will be discussed in detail (§ 12, **b**).

c. *Simple types of Moebius transformations.* If $c = 0$ in (6.1) we may assume $d = 1$ so that $\mathfrak{H}(z)$ becomes an 'integral' linear function:

$$Z = az+b.$$

If $c \neq 0$, the formula (6.1) may be written in the form

$$Z = \frac{a}{c} - \frac{\delta}{c}\frac{1}{cz+d}. \tag{6.3}$$

[1] It may be noted that the composition of functional transformations is always associative.

Thus a Moebius transformation appears as a product of Moebius transformations of the following 'simple types':

1. *Translation*

$$Z = z+b, \qquad \text{Matrix } \mathfrak{T}_b = \begin{pmatrix} 1 & b \\ 0 & 1 \end{pmatrix} \quad (b \neq 0, \text{complex}). \tag{6.31}$$

Each straight line parallel to the translation vector $\overrightarrow{0b}$ is left invariant (or transformed into itself) by the translation (6.31). The lines of every other pencil of parallel straight lines are interchanged; in particular this is true of the pencil perpendicular to $\overrightarrow{0b}$. To each of these pencils corresponds on the sphere a parabolic pencil of circles, all passing through the point **S**. This is the only point left invariant by the spherical transformation that corresponds to (6.31); hence $z = \infty$ is the only fixed point of the translation (cf. § 2, ex. 8).

The translation $\mathfrak{T}_b$ is the product (that is, the result of composition) of two inversions.

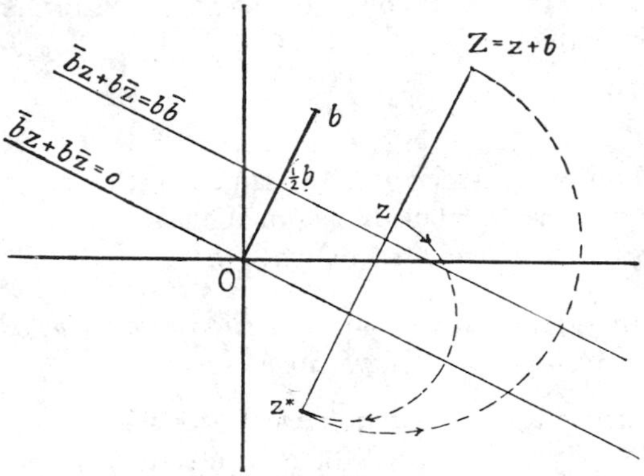

FIG. 10

Proof. Geometrical aspect suggests choosing straight lines as fundamental circles for the inversion. These lines must be perpendicular to the translation vector $\overrightarrow{0b}$. For the first inversion we take the line through 0, given by the equation $\bar{b}z+b\bar{z} = 0$. As corresponding inversion we have then

$$z^* = -\frac{b}{\bar{b}}\bar{z}.$$

As second fundamental circle we choose the line $\bar{b}z+b\bar{z} = b\bar{b}$ which has the distance $\frac{1}{2}|b|$ from 0, measured in the direction $\overrightarrow{0b}$ (cf. Figure 10). The corresponding inversion carries z^* into

$$Z = -\frac{b\overline{z^*}-b\bar{b}}{\bar{b}} = \frac{\bar{b}z+b\bar{b}}{\bar{b}} = z+b.$$

2. *Rotation* about 0 by the angle α:

$$Z = e^{i\alpha}z, \qquad \text{Matrix } \mathfrak{R}_\alpha = \begin{pmatrix} e^{i\alpha} & 0 \\ 0 & 1 \end{pmatrix}. \tag{6.32}$$

Every circle about 0 is invariant. Interchanged are the circles of the pencil that is orthogonal to the pencil of those concentric circles, that is, the pencil of all straight lines through 0. To the rotations about O in the plane correspond rotations of the sphere about the polar axis NS. If α is not an integral multiple of 2π, then N and S are the only fixed points of these spherical rotations. Accordingly 0 and ∞ are the only fixed points of the rotation (6.32).

Every rotation $\mathfrak{R}_\alpha$ is the product of two inversions.

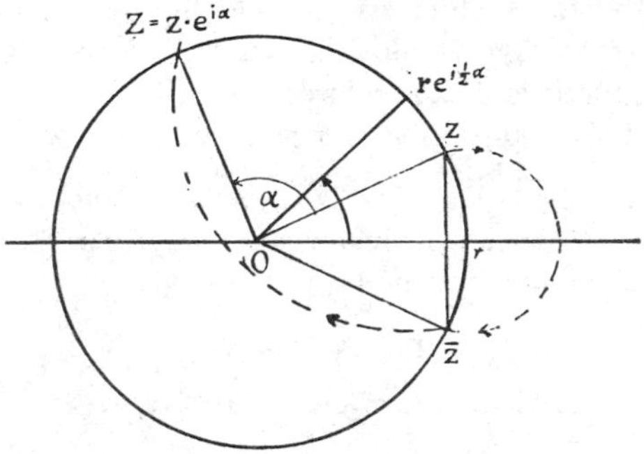

FIG. 11

Proof. As fundamental circle for the first inversion we choose the real axis: $iz - i\bar{z} = 0$. Hence $z^* = \bar{z}$. As second fundamental circle we take the line through 0 which makes the angle $\frac{1}{2}\alpha$ with the real axis:

$$ie^{-i(\alpha/2)}z - ie^{i(\alpha/2)}\bar{z} = 0.$$

The second inversion takes z^* into $Z = (e^{i(\alpha/2)}/e^{-i(\alpha/2)})\,\overline{z^*} = e^{i\alpha}z$ (Figure 11).

3. *Dilatation* about 0 with the extension coefficient $\rho > 0$:

$$Z = \rho z, \qquad \text{Matrix } \mathfrak{D}_\rho = \begin{pmatrix} \rho & 0 \\ 0 & 1 \end{pmatrix}. \tag{6.33}$$

Every line through 0 is invariant. All the circles about 0 are interchanged. Fixed points are again 0 and ∞.

Every dilatation $\mathfrak{D}_\rho$ is the product of two inversions.

Proof. As first fundamental circle take the unit circle about 0:

$$z^* = \frac{1}{\bar{z}}.$$

For the second inversion take the circle with the radius $\sqrt{\rho}$ about 0; then

$$Z = \frac{\rho}{z^*} = \rho z$$

4. *Reciprocation*

$$Z = \frac{1}{z}, \qquad \text{Matrix } \mathfrak{R} = \begin{pmatrix} 0 & 1 \\ 1 & 0 \end{pmatrix}. \tag{6.34}$$

On the sphere, reciprocation carries the point **P** (ξ, η, ζ) into the point $(\xi, -\eta, -\zeta)$; for if $z = (\xi+i\eta)/(1+\zeta)$, then $1/z = (\xi-i\eta)/(1-\zeta)$. Hence on the sphere the reciprocation is a rotation of 180° about the ξ-axis. All circles in planes parallel to the η, ζ-plane and all great circles whose planes contain the ξ-axis are invariant. Thus every circle through 1 and -1, and every circle of the orthogonal hyperbolic pencil is transformed onto itself.

From the interpretation of the reciprocation as a rotation of the sphere it is obvious that also the *reciprocation is a product of two inversions.*

d. *Mapping properties of the Moebius transformation.* The matrix notation of the simple types of Moebius transformations enables us to write the matrix of an integral Moebius transformation in the form

$$\mathfrak{H} = \mathfrak{T}_b \mathfrak{R}_\alpha \mathfrak{D}_{|a|}$$

and in virtue of (6.3) we have, if $c = |c|e^{i\gamma} \neq 0$ and $-\delta/c = |\delta/c|e^{i\phi}$, in the non-integral case,

$$\mathfrak{H} = \mathfrak{T}_{a/c} \mathfrak{D}_{|\delta/c|} \mathfrak{R}_\phi \mathfrak{R} \mathfrak{T}_d \mathfrak{D}_{|c|} \mathfrak{R}_\gamma$$

Thus every Moebius transformation appears as a product of not more than seven transformations of the four simple types in a certain order. Therefore it is a product of an even number ($\leqslant 14$) of inversions.

From the theorems of § 2, **b** we conclude

Theorem B. *Every Moebius transformation* (6.1) *is a conformal mapping of the completed z-plane onto the completed Z-plane. It carries circles into circles, real circles (including straight lines) into real circles or lines and imaginary circles into imaginary circles.*

According to § 1, **a** every circle $\mathfrak{C}$ can be given an orientation. It is geometrically evident that mapping with a Moebius transformation of one of the first three types (translation, rotation, and dilatation) preserves the orientation of every circle in the plane. Its interior is mapped into the interior of the image circle.

The geometrical construction of the point $1/z$ for a given $z \neq 0$ (cf. § 2, ex. 2) shows that reciprocation preserves the orientation of a circle which does not contain 0, but inverts the orientation of any circle containing 0 as an interior point. Hence

Corollary. *Every Moebius transformation $\mathfrak{H}$ preserves the orientation of any circle which does not contain the pole of $\mathfrak{H}$. If $\mathfrak{C}$ contains the pole of $\mathfrak{H}$, then the image circle $\mathfrak{C}_1$ of $\mathfrak{C}$ has its orientation opposite to that of $\mathfrak{C}$.*

If the pole of $\mathfrak{H}$ lies on $\mathfrak{C}$ then the image of $\mathfrak{C}$ is a straight line.

Remark. Later we shall see that a Moebius transformation can be written as a product of either two or four inversions.

Any anti-homography is obtained by following up the inversion $z^* = \bar{z}$ with respect to the real axis by a suitable Moebius transformation. Hence an anti-homography appears as a product of an odd number ($\leqslant 15$) of inversions. It is therefore an isogonal transformation inverting the sense of rotation. It carries circles into circles.

Theorem C. *Let z_1, z_2, z_3, z_4 be any four points in the completed plane no three of which are equal. Let $\mathfrak{H}$ be a Moebius transformation and $Z_j = \mathfrak{H}(z_j)$ $(j = 1, 2, 3, 4)$. Then*

$$(Z_1, Z_2; Z_3, Z_4) = (z_1, z_2; z_3, z_4).$$

More briefly: the cross ratio is an invariant ('four-point invariant') of the group of all Moebius transformations.

Proof. If the four points z_j are transformed into z_j^* by an inversion, then their cross ratio is taken into its conjugate value (cf. § 5, **c**, Theorem B); thus the cross ratio is invariant under a product of two or any even number of inversions. (Another proof will be indicated in ex. 8.)

From Theorem C it follows that

$$(Z, Z_1; Z_2, Z_3) = (z, z_1; z_2, z_3) \tag{6.35}$$

is an implicit representation of the Moebius transformation which changes z_1, z_2, z_3 into Z_1, Z_2, Z_3 respectively. Hence we have

Theorem D. *Let there be given three different points z_1, z_2, z_3 in the z-plane and three different points Z_1, Z_2, Z_3 in the Z-plane. The Moebius transformation $\mathfrak{H}$ by which*

$$Z_1 = \mathfrak{H}(z_1), \qquad Z_2 = \mathfrak{H}(z_2), \qquad Z_3 = \mathfrak{H}(z_3),$$

is then uniquely determined. It is obtained by solving the equation (6.35) *with respect to Z.*

Putting $Z_j = z_j$ $(j = 1, 2, 3)$ we conclude that *a Moebius transformation with three distinct fixed points is necessarily the identity.* This is an alternative to the uniqueness statement of Theorem D. For if $\mathfrak{H}_1, \mathfrak{H}_2$ are two Moebius transformations such that $Z_j = \mathfrak{H}_1(z_j) = \mathfrak{H}_2(z_j)$, then $\mathfrak{H}_2^{-1}\mathfrak{H}_1$ is a Moebius transformation with three distinct fixed points, that is, the identity; therefore the transformations $\mathfrak{H}_1, \mathfrak{H}_2$ are identical. (Cf. Appendix 1.)

From Theorem D follows another proof of Theorem A in § 5, **c**. Observe that three different points always determine uniquely the real circle $\mathfrak{C}$ through these points. Now prescribe three distinct points X_1, X_2, X_3 on the real axis as images of the points z_1, z_2, z_3; the Moebius transformation uniquely defined by

$$(Z, X_1; X_2, X_3) = (z, z_1; z_2, z_3)$$

will transform the circle $\mathfrak{C}$ into the real axis which means that Z will be real if and only if z is a point on the circle $\mathfrak{C}$. Since the cross ratio of any four points on the real axis is obviously real, we conclude that the cross ratio of four points in the completed plane is real if and only if they lie on a circle (Cf. also § 1, ex. 9).

The preceding discussion may be completed by the

Corollary. *Every transformation $Z = f(z)$ of the completed z-plane onto itself, that has the cross ratio as a four-point invariant, is necessarily a Moebius transformation.*

This follows immediately from the fact that by solving the equation (6.35) with respect to Z one obtains a Moebius transformation.

e. *Transformation of a circle.* Let a circle be given by its hermitian matrix $\mathfrak{C} = \begin{pmatrix} A & B \\ C & D \end{pmatrix}$, and thus by its equation (1.2). By a Moebius transformation $Z = \mathfrak{H}(z)$ the circle $\mathfrak{C}$ is mapped onto another circle which may be denoted symbolically by $\mathfrak{H}(\mathfrak{C})$. Let $\mathfrak{C}_1 = \begin{pmatrix} A_1 & B_1 \\ C_1 & D_1 \end{pmatrix}$ be its hermitian matrix. Its equation is therefore

$$\mathfrak{C}_1(Z, \bar{Z}) = A_1 Z\bar{Z} + B_1 Z + C_1 \bar{Z} + D_1 = 0. \tag{6.4}$$

What is the connection between the hermitian matrices $\mathfrak{C}$ and $\mathfrak{C}_1$?

This question will most easily be answered by making use of homogeneous variables (cf. § 6, **a**). The equation (1.2) then may be written in the form

$$z_2\bar{z}_2\mathfrak{C}(z, \bar{z}) = Az_1\bar{z}_1 + Bz_1\bar{z}_2 + C\bar{z}_1 z_2 + Dz_2\bar{z}_2 = 0$$

or, in matrix notation

$$\mathfrak{z}'\mathfrak{C}\bar{\mathfrak{z}} = 0, \qquad \mathfrak{z} = \begin{pmatrix} z_1 \\ z_2 \end{pmatrix}. \tag{6.41}$$

Now if $Z = \mathfrak{H}(z)$, then by (6.12) in homogeneous coordinates

$$\mathfrak{z} = \frac{1}{q}\mathfrak{H}^{-1}\mathfrak{Z} = \frac{1}{q_1}\mathfrak{S}\mathfrak{Z}, \quad \mathfrak{S} = \begin{pmatrix} d & -b \\ -c & a \end{pmatrix} = \delta\mathfrak{H}^{-1}$$

and the equation (6.41) will become

$$\mathfrak{Z}'\mathfrak{S}'\mathfrak{C}\bar{\mathfrak{S}}\bar{\mathfrak{Z}} = 0.$$

Thus we have for the transformed circle the equation

$$\mathfrak{C}_1(Z, \bar{Z}) = 0 \qquad \text{where } \mathfrak{C}_1 = \mathfrak{S}'\mathfrak{C}\bar{\mathfrak{S}}. \tag{6.42}$$

This matrix is hermitian: $\mathfrak{C}_1' = \bar{\mathfrak{S}}'\mathfrak{C}'\mathfrak{S} = \bar{\mathfrak{S}}'\bar{\mathfrak{C}}\mathfrak{S} = \bar{\mathfrak{C}}_1$, and since

$$|\mathfrak{C}_1| = |\mathfrak{S}||\mathfrak{C}||\bar{\mathfrak{S}}| = |\det \mathfrak{S}|^2|\mathfrak{C}|,$$

the mapping of $\mathfrak{C}$ onto $\mathfrak{C}_1$ does not change the sign of the discriminant. This provides another proof of the statements concerning the transformation of circles in Theorem B.

Every real circle $\mathfrak{C}$ may be transformed into the real unit circle $z\bar{z}-1 = 0$. Thus there is a Moebius transformation $\mathfrak{S}$ such that

(6.43)
$$\mathfrak{S}'\mathfrak{C}\bar{\mathfrak{S}} = \begin{pmatrix} 1 & 0 \\ 0 & -1 \end{pmatrix}.$$

If the circle $\mathfrak{C}$ is imaginary (that is, the hermitian matrix $\mathfrak{C}$ positive definite) it is known that there is a matrix $\mathfrak{S}$ such that

(6.44)
$$\mathfrak{S}'\mathfrak{C}\bar{\mathfrak{S}} = \mathfrak{E}.$$

Hence every imaginary circle may be transformed onto the 'imaginary unit circle' $z\bar{z}+1 = 0$ by a suitable Moebius transformation. This result will be proved in another way in § 9, **d**.

f. *Involutions.* A transformation (or function) $Z = f(z)$ which is not the identity ($Z = z$) is said to be involutory if $f(f(z)) = z$ for all z (cf. § 2, **a**). An involutory Moebius transformation will be called an *involution*. The matrix

$$\mathfrak{H} = \begin{pmatrix} a & b \\ c & d \end{pmatrix}$$

of an involution has to satisfy the condition

(6.5)
$$\mathfrak{H}^2 = \mu\mathfrak{E} = \begin{pmatrix} \mu & 0 \\ 0 & \mu \end{pmatrix} \qquad (\mu \neq 0, \text{ complex}).$$

By working out the matrix $\mathfrak{H}^2$ we obtain $\begin{pmatrix} a^2+bc & (a+d)b \\ (a+d)c & bc+d^2 \end{pmatrix}$; thus

Theorem E. *The Moebius transformation* (6.1) *is an involution if and only if*

$$a+d = 0.$$

It may be noted that this relation expresses the fact that the bilinear function on the left-hand side of (6.13) is symmetric in z and Z, and this is the necessary and sufficient condition for the corresponding Moebius transformation $\mathfrak{H}$ to coincide with its own inverse, that is, to be an involution.

Every involution $\mathfrak{H}$ has evidently the following property. For any point z_1 of the completed plane let $\mathfrak{H}(z_1) = z_2$; then also $\mathfrak{H}(z_2) = z_1$. Two points z_1, z_2 related in this way are said to form a 'conjugate pair' with respect to the involution $\mathfrak{H}$. Every point in the plane is the element of a conjugate pair.

Theorem F. *If the Moebius transformation* $Z = \mathfrak{H}(z)$ *possesses one conjugate pair* z_1, z_2 $(z_1 \neq z_2)$ *such that* $z_2 = \mathfrak{H}(z_1)$ *and* $z_1 = \mathfrak{H}(z_2)$, *then* $\mathfrak{H}$ *is an involution.*

Proof. Let z_3 be different from z_1 and from z_2 and let $\mathfrak{H}(z_3) = Z_3$. Then by (6.35)

$$(Z, z_2; z_1, Z_3) = (z, z_1; z_2, z_3).$$

Further, if $\mathfrak{H}(Z_3) = z_3'$, then

$$(Z_3, z_1; z_2, z_3) = (z_3', z_2; z_1, Z_3)$$
$$= (Z_3, z_1; z_2, z_3') \qquad \text{by (5.41).}$$

Hence $z_3' = z_3$, q.e.d.

Every involution $\mathfrak{H} = \begin{pmatrix} a & b \\ c & -a \end{pmatrix}$ has two distinct fixed points γ_1, γ_2, the roots of the equation $\mathfrak{H}(z) = z$, that is, the quadratic equation

$$cz^2 - 2az - b = 0, \tag{6.51}$$

whose discriminant, $-4|\mathfrak{H}|$, is not zero. If c is non-zero, both fixed points are finite. Since

$$\gamma_1 + \gamma_2 = 2\frac{a}{c} = 2z_\infty \tag{6.52}$$

it follows that the pole z_∞ of $\mathfrak{H}$ lies midway between γ_1 and γ_2. Therefore the straight line I joining the fixed points is mapped onto itself by the involution $\mathfrak{H}$.

If $c = 0$ we may take $d = 1$; thus $Z = -z + b$ is the most general integral involution. One of its fixed points coincides with the pole at infinity.

Finally we notice that there is one and only one involution which has a given pair of points γ_1, γ_2 as fixed points. Cf. ex. 9.

EXAMPLES

1. *Interchangeability* (commutativity) of simple types.

$$\mathfrak{T}_a\mathfrak{T}_b = \mathfrak{T}_b\mathfrak{T}_a, \qquad \mathfrak{R}_\alpha\mathfrak{R}_\beta = \mathfrak{R}_\beta\mathfrak{R}_\alpha, \qquad \mathfrak{D}_\rho\mathfrak{D}_\sigma = \mathfrak{D}_\sigma\mathfrak{D}_\rho, \qquad \mathfrak{R}_\alpha\mathfrak{D}_\rho = \mathfrak{D}_\rho\mathfrak{R}_\alpha.$$

Show that if $b \neq 0$ and α is not an integral multiple of 2π, then

$$\mathfrak{T}_b\mathfrak{R}_\alpha \neq \mathfrak{R}_\alpha\mathfrak{T}_b.$$

Thus, like matrix multiplication, the composition (multiplication) of Moebius transformations is *not commutative* in general. Also discuss the products

$$\mathfrak{T}_b\mathfrak{D}_\rho, \qquad \mathfrak{R}\mathfrak{T}_b, \qquad \mathfrak{R}\mathfrak{R}_\alpha, \qquad \mathfrak{R}\mathfrak{D}_\rho$$

as to interchangeability of the factors (cf. also § 8, ex. 8).

2. Find the matrix $\mathfrak{H}$ of the Moebius transformation which represents the rotation by the angle α about the point γ in the complex plane.

3. Show that every translation $\mathfrak{T}_b$ can be written as the product of two rotations about suitably chosen centres γ_1, γ_2.

(The angles of the two rotations must be opposite, equal, and not integral multiples of 2π; after a choice of the angle has been made, the centre γ_1 may be chosen *ad libitum*, and then γ_2 will be uniquely defined. Observe that the rotation factors are not commutative in general.)

4. All integral Moebius transformations (similarities) form a group. The ratio $(z_1; z_2, z_3)$ is a three-point invariant of this group. Conversely: every transformation $Z = f(z)$ for which $(z_1; z_2, z_3)$ is invariant, is an integral Moebius transformation $Z = az+b$.

5. All displacements (or motions) $Z = e^{i\alpha}z+b$ (α real) form a subgroup of the group of all integral Moebius transformations. For the subgroup the distance $|z_1-z_2|$ is a two-point invariant.

6. All rotations about 0 form a group for which $|z|$ is a one-point invariant.

7. All translations $\mathfrak{T}_b$ form a group for which the difference z_1-z_2 is a two-point invariant.

8. Theorem C may be proved by making use of the decomposition of a Moebius transformation $\mathfrak{H}$ into simple type factors (cf. **d**). The first three of the simple types do not even change the ratio of three points. As to the reciprocation, if none of the four points z_1, z_2, z_3, z_4 is zero, we have

$$\left(\frac{1}{z_1}; \frac{1}{z_3}, \frac{1}{z_4}\right) = \frac{z_4}{z_3}(z_1; z_3, z_4), \qquad \left(\frac{1}{z_2}; \frac{1}{z_3}, \frac{1}{z_4}\right) = \frac{z_4}{z_3}(z_2; z_3, z_4)$$

and therefore
$$\left(\frac{1}{z_1}, \frac{1}{z_2}; \frac{1}{z_3}, \frac{1}{z_4}\right) = (z_1, z_2; z_3, z_4).$$

If for instance $z_3 = 0$, by (5.43)

$$\left(\frac{1}{z_1}, \frac{1}{z_2}; \infty, \frac{1}{z_4}\right) = \left(\frac{1}{z_4}; \frac{1}{z_2}, \frac{1}{z_1}\right) = \frac{z_1}{z_2}(z_4; z_2, z_1) = (z_1, z_2; 0, z_4).$$

9. Let a Moebius transformation be given implicitly by the equation

$$(Z, z; \gamma_1, \gamma_2) = k \qquad (\gamma_2 \neq \gamma_1)$$

where k is a complex constant, not zero. Show that γ_1, γ_2 are the fixed points (§ 2, ex. 8) of this transformation. For what value of k will this Moebius transformation be an involution?

10. Prove Theorem F without making use of cross ratios.

(If z_1 is not a fixed point of $\mathfrak{H}$, that is,

$$cz_1^2-(a-d)z_1-b \neq 0,$$

then from $\mathfrak{H}(\mathfrak{H}(z_1)) = z_1$ it follows that $a+d = 0$.)

11. Every real circle $\mathfrak{C}$ is a Moebius image of the real axis and its points z can therefore be represented in the form $z = \mathfrak{H}(t)$, where t is a real parameter (cf. § 1, ex. 9). Obtain a parametric representation of this kind for the real unit circle. Interpret the result in the sense of § 6, **e**.

12. Every anti-homography transforms four points with the cross ratio λ into four points with the cross ratio $\bar{\lambda}$.

13. The theorem of § 2, ex. 1 may be extended to the case of an imaginary $\mathfrak{C}$ and inversion with respect to a real circle $\mathfrak{C}_0$. Since every real circle may be mapped by a Moebius transformation onto the real unit circle, it will be sufficient to prove the theorem for this circle in place of $\mathfrak{C}_0$. The inversion $z^* = 1/\bar{z}$ then leaves invariant every circle $\mathfrak{C}$ orthogonal to the unit circle: $\mathfrak{C} = \begin{pmatrix} A & B \\ \bar{B} & A \end{pmatrix}$, and every invariant circle $\mathfrak{C}$ has a hermitian matrix with the two diagonal elements equal to each other. Taking into account the earlier result and § 1, ex. 11, we have thus the general theorem: *A circle $\mathfrak{C} \neq \mathfrak{C}_0$ is mapped onto itself by inversion with respect to $\mathfrak{C}_0$ if and only if $\mathfrak{C}$ is orthogonal to $\mathfrak{C}_0$.*

14. Generalizing the final remark of § 6, **f**, it can be shown that an involution is uniquely defined if two couples of conjugate points z_1, Z_1 and z_2, Z_2 are prescribed arbitrarily. Either z_1 or z_2 or both may be fixed points. If all the four points are different, then the fixed points of the involution defined by the two couples form a harmonic set with each of the couples. The involution is given by

$$Z = \frac{(z_1Z_1 - z_2Z_2)z + z_2Z_2(z_1 + Z_1) - z_1Z_1(z_2 + Z_2)}{(z_1 + Z_1 - z_2 - Z_2)z + z_2Z_2 - z_1Z_1}.$$

15. Making use of the algebraic method of § 6, **e** show that a Moebius transformation $\mathfrak{H}$ inverts the orientation of a circle $\mathfrak{C}$ which has its centre at the pole of $\mathfrak{H}$. (Cf. § 1, **a**.)

§ 7. Real one-dimensional projectivities

a. *Perspectivities.* The points z and $\tilde{z}$ on two different straight lines $\mathfrak{l}$ and $\tilde{\mathfrak{l}}$ are said to lie *in perspective*, and the mapping that leads from $\mathfrak{l}$ to $\tilde{\mathfrak{l}}$ is called a *perspectivity*, if there is a point z_0, which lies neither on $\mathfrak{l}$ nor on $\tilde{\mathfrak{l}}$, such that for every z on $\mathfrak{l}$ there is a straight line containing z, z_0, $\tilde{z}$. The point z_0 is called the *centre* of the perspectivity.

Theorem A. *Every perspectivity can be represented by a Moebius transformation possessing two different fixed points.*

Proof. Any line of the pencil of straight lines through z_0 is given by an equation

$$cz + \bar{c}\bar{z} = cz_0 + \bar{c}\bar{z}_0 \tag{7.1}$$

where $c \neq 0$ is a complex parameter. A line of this pencil is therefore defined by

$$w = \frac{\bar{c}}{c} \qquad (|w| = 1).$$

Let the line I be given by

$$az+\bar{a}\bar{z} = a_1 \qquad (a \neq 0, \quad a_1 \text{ real}).$$

The pencil line (7.1) will intersect this line at

$$z = \frac{a_1\bar{c}-\bar{a}(cz_0+\bar{c}\bar{z}_0)}{a\bar{c}-\bar{a}c} = \frac{(a_1-\bar{a}\bar{z}_0)w-\bar{a}z_0}{aw-\bar{a}} = \mathfrak{H}_a(w).$$

The matrix

$$\mathfrak{H}_a = \begin{pmatrix} a_1-\bar{a}\bar{z}_0 & -\bar{a}z_0 \\ a & -\bar{a} \end{pmatrix} \tag{7.11}$$

has the determinant $|\mathfrak{H}_a| = \bar{a}(az_0+\bar{a}\bar{z}_0-a_1) \neq 0$ since z_0 is not a point of I.

If the line $\tilde{\text{I}}$ has the equation $bz+\bar{b}\bar{z} = b_1$ ($b \neq 0$, b_1 real, $bz_0+\bar{b}\bar{z}_0 \neq b_1$), then the line (7.1) meets I at the point

$$\tilde{z} = \mathfrak{H}_b(w)$$

where the matrix $\mathfrak{H}_b$ is formed according to (7.11). Hence z, z_0, $\tilde{z}$ are collinear and

$$\tilde{z} = \mathfrak{H}(z) \quad \text{with} \quad \mathfrak{H} = \mathfrak{H}_b\mathfrak{H}_a^{-1} \tag{7.12}$$

represents the perspectivity.

One fixed point of $\mathfrak{H}$ is obviously the common point of I and $\tilde{\text{I}}$, that is,

$$Z_0 = \frac{a_1\bar{b}-b_1\bar{a}}{a\bar{b}-b\bar{a}}. \tag{7.13}$$

The other fixed point is the centre z_0 of the perspectivity:

$$\mathfrak{H}_a^{-1}(z_0) = 0, \qquad \mathfrak{H}_b(0) = z_0,$$

hence by (7.12)

$$\mathfrak{H}(z_0) = z_0.$$

Thus the theorem is proved.

Let us denote by z_∞ the pole of $\mathfrak{H}$ so that $\mathfrak{H}(z_\infty) = \infty$, and by Z_∞ the pole of $\mathfrak{H}^{-1}$ so that $\mathfrak{H}(\infty) = Z_\infty$. Then z_∞ is the one point in the plane which is carried by $\mathfrak{H}$ into infinity and since the image of I is again a line, viz. $\tilde{\text{I}}$, it follows that z_∞ is a point on I. In the same way it is shown that Z_∞ is a point on $\tilde{\text{I}}$ (cf. Figure 12). Because the perspectivity from the centre z_0 also transforms z_∞ into ∞ and ∞ into Z_∞, we conclude that the line I* through z_0 and z_∞ is parallel to $\tilde{\text{I}}$, and the line $\tilde{\text{I}}$* through z_0 and Z_∞ parallel to I. Hence

Corollary. *If the Moebius transformation $\mathfrak{H}$ represents a perspectivity, then its fixed points z_0, Z_0 and the poles z_∞, Z_∞ of $\mathfrak{H}$, $\mathfrak{H}^{-1}$ respectively are pairs of opposite vertices of a parallelogram which will be called the* characteristic parallelogram *of $\mathfrak{H}$.*

The Moebius transformation $\mathfrak{H}$ is representative of another perspectivity; its centre is Z_0. It transforms the line $\mathfrak{l}^*$ into the line $\tilde{\mathfrak{l}}^*$.

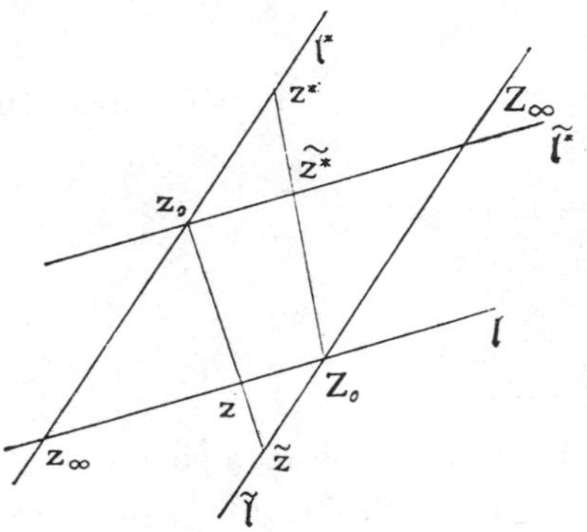

FIG. 12

b. *Projectivities.* A *projectivity* between two straight lines, equal or different, is, by definition, a product of one or several perspectivities with equal or different centres. From the preceding discussion we conclude that *every projectivity between equal or different straight lines* $\mathfrak{l}$, $\tilde{\mathfrak{l}}$ *in the complex plane can be represented by a Moebius transformation.*

Sets of points on two (equal or different) lines which can be transformed into one another by a projectivity are said to be projective. From § 6, Theorem C it follows that two projective quadruplets of points on equal or different lines have equal cross ratios.

A simple geometrical consideration shows that any three points z_1, z_2, z_3 on a line $\mathfrak{l}$ can be carried into any three points $\tilde{z}_1$, $\tilde{z}_2$, $\tilde{z}_3$ on another line $\tilde{\mathfrak{l}}$ by a succession (product) of not more than two perspectivities (cf. Figure 13). Draw the line $\mathfrak{l}^*$ joining z_1 and $\tilde{z}_2$, and the line $\mathfrak{l}_1$ joining z_1 and $\tilde{z}_1$, and the line $\mathfrak{l}_2$ joining z_2 and $\tilde{z}_2$. Join the intersection z_0^* of $\mathfrak{l}_1$ and $\mathfrak{l}_2$ by a line with $\tilde{z}_3$; this line meets $\mathfrak{l}^*$ in z_3^*. Draw the line through z_3 and z_3^*; it meets $\mathfrak{l}_2$ at z_0. Then z_0 is the centre of the perspectivity carrying z_1, z_2, z_3 into z_1, $\tilde{z}_2$, z_3^* on $\mathfrak{l}^*$; and z_0^* the centre of the perspectivity carrying these three points into $\tilde{z}_1$, $\tilde{z}_2$, $\tilde{z}_3$ on $\tilde{\mathfrak{l}}$.

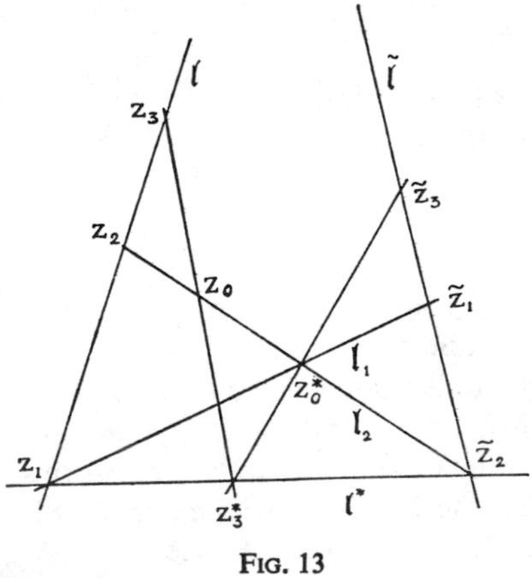

FIG. 13

By a further projectivity it will be possible to map the three points on $\tilde{\mathfrak{l}}$ onto any three points on $\mathfrak{l}$. Hence by a succession of not more than four perspectivities any three points on a line can be carried into any three points on the same line.

By § 6, Theorem D there is one and only one Moebius transformation, by means of which this operation can be performed. Since every projectivity can be represented by a Moebius transformation, we conclude

Theorem B. *There is one and only one projectivity which maps three points on a line onto three given points on the same or another line.*

As far as projective mapping on one and the same line is concerned, this is the so-called *fundamental theorem of real one-dimensional projective geometry.* Indeed the straight line in our geometry of complex numbers, also called conformal geometry, is the (closed) real projective line, namely a circle. As to the mapping from one line onto the same or another line, the theorem is concerned with transformations within the parabolic bundle of all circles whose common point is at infinity (cf. § 4, b, (ii)).

Theorem C. *A Moebius transformation* $\mathfrak{H}$ *represents a projectivity of the points of any line* $\mathfrak{l}$ *through its pole* z_∞ *onto a line* $\tilde{\mathfrak{l}}$ *through the pole* Z_∞ *of* $\mathfrak{H}^{-1}$. *If* $\mathfrak{l}$ *passes through the fixed point* Z_0 *(or* z_0*), then* $\tilde{\mathfrak{l}}$ *passes through the same point; if* $\tilde{\mathfrak{l}}$ *does not coincide with the line* $\mathfrak{l}$ *through* Z_0 *(or* z_0*), then the projectivity* $\mathfrak{l} \to \tilde{\mathfrak{l}}$ *is the perspectivity whose centre is the other fixed point.*

Proof. If $\mathfrak{H}$ is an integral Moebius transformation, then one of its fixed points coincides with the pole at infinity, and $\mathfrak{l}$ as well as its image $\tilde{\mathfrak{l}}$ must pass through the finite fixed point if it exists. If $\mathfrak{H}$ is a translation (both fixed points coincide at infinity) then every line can be chosen as $\mathfrak{l}$, and $\tilde{\mathfrak{l}}$ is parallel to $\mathfrak{l}$. In any case the projectivity will be a perspectivity with centre at infinity.

If, however, in the matrix $\mathfrak{H} = \begin{pmatrix} a & b \\ c & d \end{pmatrix}$ the coefficient $c \neq 0$, then the fixed points z_0, Z_0 of $\mathfrak{H}$ are finite (different or equal). Every line $\mathfrak{l}$ through z_∞ is mapped by $\mathfrak{H}$ onto a line $\tilde{\mathfrak{l}}$ through Z_∞ and the mapping represents a projectivity $\mathfrak{l} \to \tilde{\mathfrak{l}}$. If the fixed points are different and the pole z_∞ is not collinear with the fixed points, then the line $\mathfrak{l}$ through Z_0 and z_∞ is mapped by $\mathfrak{H}$ onto the line $\tilde{\mathfrak{l}}$ through Z_0 and the pole Z_∞ of $\mathfrak{H}^{-1}$. This projectivity, having the common point Z_0 of $\mathfrak{l}$ and $\tilde{\mathfrak{l}}$ as fixed point, is then a perspectivity whose centre is the second fixed point z_0 of $\mathfrak{H}$. Thus we have again the characteristic parallelogram with the vertices z_0, Z_0, z_∞, Z_∞ (cf. the Corollary of § 7, **a**, and § 8, **e**).

Further, it is evident that if z_0, Z_0, z_∞ are collinear, than Z_∞ lies on the common line of these points (cf. ex. 5) and this line is mapped onto itself by $\mathfrak{H}$. The mapping can be effected by a succession of not more than three perspectivities.[2]

If in particular $Z_\infty = z_\infty$, then $\mathfrak{H}$ is an *involution* and therefore cannot represent a perspectivity between two lines (cf. ex. 1). Assuming $c \neq 0$ we shall construct *two perspectivities* $\mathfrak{A}$, $\mathfrak{B}$ *such that* $\mathfrak{H} = \mathfrak{B}\mathfrak{A}$. As centre of $\mathfrak{A}$ we choose a point A not on the line $\mathfrak{l}$ through z_0 and Z_0, from which we project onto a line $\mathfrak{l}'$ through z_0 parallel to $\overrightarrow{AZ_0}$ (cf. Figure 14). Then

$$\mathfrak{A}(\infty) = Z'_\infty, \qquad \mathfrak{A}(z_0) = z_0, \qquad \mathfrak{A}(z_\infty) = z'_\infty, \qquad \mathfrak{A}(Z_0) = \infty.$$

[2] Cf. Coxeter [2], p. 39, §4·22

As centre of $\mathfrak{B}$ we choose the point B on the line through A and Z_0, having from Z_0 the same distance as A; then cf. (6.52)

$$\mathfrak{B}(Z'_\infty) = z_\infty, \qquad \mathfrak{B}(z_0) = z_0, \qquad \mathfrak{B}(z'_\infty) = \infty, \qquad \mathfrak{B}(\infty) = Z_0.$$

Therefore

$$\mathfrak{B}[\mathfrak{A}(\infty)] = z_\infty, \qquad \mathfrak{B}[\mathfrak{A}(z_0)] = z_0, \qquad \mathfrak{B}[\mathfrak{A}(z_\infty)] = \infty, \qquad \mathfrak{B}[\mathfrak{A}(Z_0)] = Z_0.$$

Therefore

$$\mathfrak{B}[\mathfrak{A}(z)] = \mathfrak{H}(z).$$

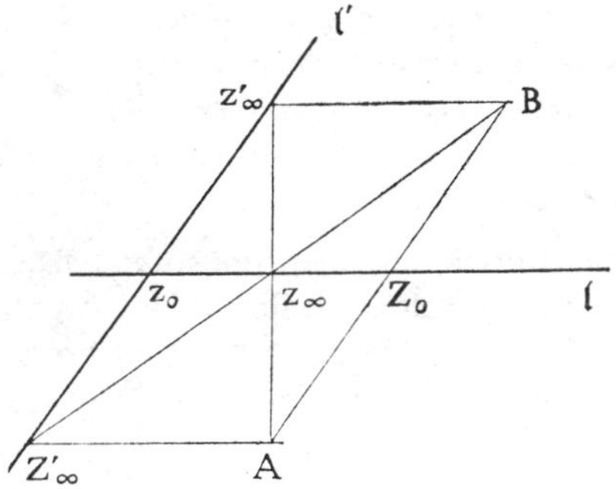

FIG. 14

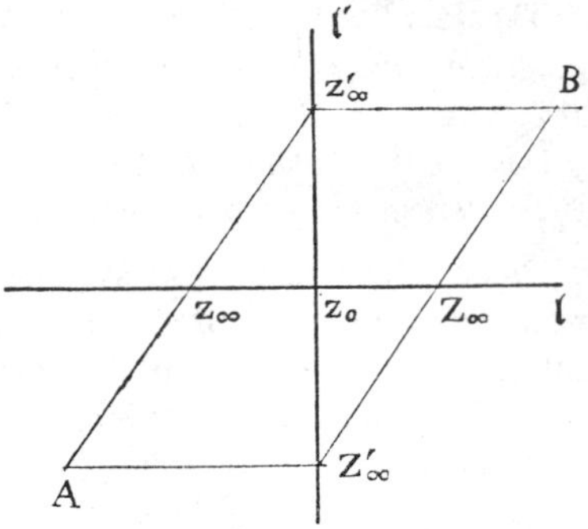

FIG. 15

Finally we have the case of a *unique finite fixed point* z_0 of the Moebius transformation $\mathfrak{H}$. This point is then the midpoint between the poles z_∞, Z_∞ (cf. ex. 6). Again two perspectivities can be found whose representing Moebius transformations $\mathfrak{A}$ and $\mathfrak{B}$ have $\mathfrak{H}$ as product. We chose an intermediate line $\mathfrak{l}'$ through z_0, different from the line $\mathfrak{l}$ through the poles. We select a point A, not on $\mathfrak{l}$ and not on $\mathfrak{l}'$, as centre for the first perspectivity $\mathfrak{A}$. Then (cf. Figure 15)

$$\mathfrak{A}(z_\infty) = z'_\infty, \qquad \mathfrak{A}(z_0) = z_0, \qquad \mathfrak{A}(\infty) = Z'_\infty.$$

For the second perspectivity $\mathfrak{B}$ we choose the centre B opposite to A on the parallelogram defined by A, z'_∞, Z'_∞; then

$$\mathfrak{B}(z'_\infty) = \infty, \qquad \mathfrak{B}(z_0) = z_0, \qquad \mathfrak{B}(Z'_\infty) = Z_\infty.$$

Thus the product $\mathfrak{B}\mathfrak{A}$ has the characteristic properties of $\mathfrak{H}$.
Hence

$$\mathfrak{B}[\mathfrak{A}(z)] = \mathfrak{H}(z).$$

c. *Line—circle perspectivity.* Let four lines $\mathfrak{l}_j$ $(j = 1, 2, 3, 4)$ through the point Z_0 be given by the equations

$$c_j z + \bar{c}_j \bar{z} = c_j Z_0 + \bar{c}_j \bar{Z}_0. \tag{7.2}$$

If no three of them coincide we define their cross ratio by

$$(\mathfrak{l}_1, \mathfrak{l}_2; \mathfrak{l}_3, \mathfrak{l}_4) = (w_1, w_2; w_3, w_4), \qquad w_j = \frac{\bar{c}_j}{c_j}, \quad (j = 1, 2, 3, 4)$$

Since $|w_j| = 1$ this cross ratio is real.

Now suppose that the line $\mathfrak{l}$ given by

$$(7.21) \qquad az + \bar{a}\bar{z} = a_1 \qquad (a \neq 0, \quad a_1 \text{ real})$$

does not pass through Z_0 and meets $\mathfrak{l}_j$ at the point z_j. It has been shown in § 7, a that there is a Moebius transformation $\mathfrak{H}_a$ such that $\mathfrak{H}_a(w_j) = z_j$. Hence

$$(7.22) \qquad (\mathfrak{l}_1, \mathfrak{l}_2; \mathfrak{l}_3, \mathfrak{l}_4) = (z_1, z_2; z_3, z_4)$$

independent of the choice of $\mathfrak{l}$.

This relation can be extended. Let Z_0 be a point on the real circle $\mathfrak{C} = (\gamma, \rho)$ and let the lines (7.2) intersect $\mathfrak{C}$ at the points Z_j then

$$(7.3) \qquad (\mathfrak{l}_1, \mathfrak{l}_2; \mathfrak{l}_3, \mathfrak{l}_4) = (Z_1, Z_2; Z_3, Z_4),$$

for any circle $\mathfrak{C}$ through Z_0.

To prove this we shall construct a Moebius transformation $\mathfrak{H}$ representing the perspectivity from $\mathfrak{l}$ to $\mathfrak{C}$ with Z_0 as centre. Let

$$(7.31) \qquad cz + \bar{c}\bar{z} = cZ_0 + \bar{c}\bar{Z}_0$$

be any line through Z_0. Then there is a Moebius transformation $\mathfrak{H}_a$ such that $z = \mathfrak{H}_a(w)$, $(w = \bar{c}/c)$ is the point where the line (7.31) meets the line $\mathfrak{l}$.

The point Z where the line (7.31) meets the circle $\mathfrak{C}$ the second time will be found as follows. Since

$$\rho = |Z_0 - \gamma|,$$

the equation of $\mathfrak{C}$, viz. $(Z-\gamma)(\bar{Z}-\bar{\gamma}) = \rho^2$, becomes

$$(Z-Z_0)(\bar{Z}-\bar{Z}_0) - (\bar{\gamma}-\bar{Z}_0)(Z-Z_0) - (\gamma - Z_0)(\bar{Z}-\bar{Z}_0) = 0.$$

By (7.31)

$$\bar{Z}-\bar{Z}_0 = -\frac{c}{\bar{c}}(Z-Z_0)$$

and therefore for $Z \neq Z_0$, we have

$$(Z-Z_0) + w(\bar{\gamma}-\bar{Z}_0) - (\gamma - Z_0) = 0$$

so that

$$Z = \gamma + \mathfrak{H}_a^{-1}(z)(\bar{Z}_0 - \bar{\gamma})$$

is the Moebius transformation representing the perspectivity from $\mathfrak{l}$ to $\mathfrak{C}$ with Z_0 as centre.

The nature of this transformation depends on the relative positions of $\mathfrak{l}$ and $\mathfrak{C}$. It will be studied in ex. 4.

EXAMPLES

1. The fact that a perspectivity can never be represented by an involution (where $Z_\infty = z_\infty$) also becomes evident if we observe that the trace of the matrix $\mathfrak{H}_b\mathfrak{H}_a^{-1}$ (cf. proof of Theorem A), viz.

$$(\bar{a}b+\bar{b}a)z_0+2\bar{a}\bar{b}\bar{z}_0-(a_1\bar{b}+b_1\bar{a})$$

can never be zero (cf. § 6, Theorem E) except when $z_0 = Z_0$ [cf. (7.13)], which is impossible.

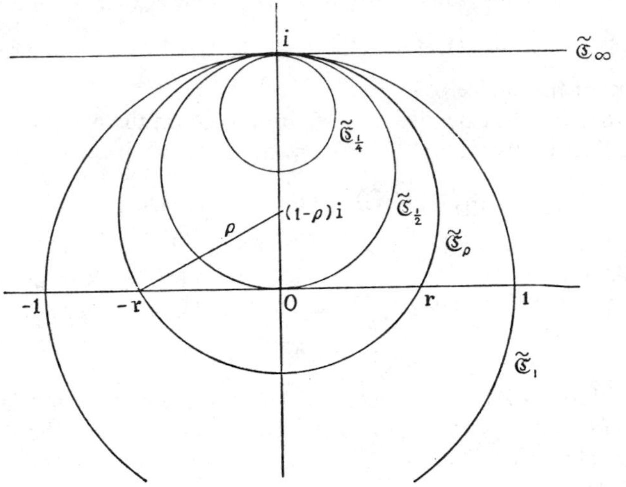

FIG. 16

2. The Moebius transformation $\mathfrak{H}$ with the two different finite fixed points z_0 and Z_0, and the pole z_∞ is implicitly given by

$$(Z, \infty; z_0, Z_0) = (z, z_\infty; z_0, Z_0).$$

Hence

$$(Z_\infty, \infty; z_0, Z_0) = (\infty, z_\infty; z_0, Z_0)$$

which is equivalent to the relation

$$Z_\infty - Z_0 = -(z_\infty - z_0)$$

expressing that z_0, Z_0 and z_∞, Z_∞ are the two pairs of opposite vertices of a parallelogram. This is the characteristic parallelogram; it was first mentioned by Jacobsthal [3] (cf. § 7, **a**).

3. Let $\mathfrak{C}_0$ be the circle with radius $\rho > 0$ about z_0. The straight line (7.1) intersects $\mathfrak{C}_0$ in two points, z and $\tilde{z} = 2z_0 - z$ which satisfy the quadratic equation

$$(z-z_0)^2+w\rho^2 = 0 \quad \text{if} \quad w = \frac{\bar{c}}{c}.$$

The four lines $\mathfrak{l}_j$ through z_0 which meet $\mathfrak{C}_0$ at z_j respectively ($j = 1, 2, 3, 4$) then have the cross ratio

$$(\mathfrak{l}_1, \mathfrak{l}_2; \mathfrak{l}_3, \mathfrak{l}_4) = ((z_1-z_0)^2, (z_2-z_0)^2; (z_3-z_0)^2, (z_4-z_0)^2)$$
$$= (z_1, z_2; z_3, z_4)(z_1, z_2; \tilde{z}_3, \tilde{z}_4).$$

4. Let $\mathfrak{C}_\rho$ be the circle

$$(Z-(1-\rho)i)(\bar{Z}+(1-\rho)i) = \rho^2 \qquad (-\infty < \rho < \infty).$$

All these $\mathfrak{C}_\rho$ form the parabolic pencil of all circles passing through $Z_0 = i$ with centres $\gamma = (1-\rho)i$ on the imaginary axis. We determine the Moebius transformation $\mathfrak{H}_\rho$ representing the perspective mapping of the real axis ($\mathfrak{l}$) onto the circle $\mathfrak{C}_\rho$ (cf. § 7, **c**) for the various values of ρ. First let $\rho > \frac{1}{2}$ so that the circle $\mathfrak{C}_\rho$ intersects the real axis at the two points $\pm r$ where

$$r = \sqrt{(2\rho-1)}, \qquad \text{(cf. Figure 16).}$$

These are then the fixed points of the Moebius transformation $\mathfrak{H}_\rho$ which is defined by the equation

$$(Z, i; -r, r) = (z, \infty; -r, r)$$

whence

$$Z = \mathfrak{H}_\rho(z) = \frac{iz+r^2}{z+i}. \tag{7.4}$$

Actually, however, this is the required Moebius transformation for all values of ρ. Here r and $-r$ are the fixed points, and i is the pole of $\mathfrak{H}_\rho^{-1}$ whatever real value ρ may have. Moreover

$$z = x \text{ (real)}, \qquad Z = \frac{ix+r^2}{x+i}, \qquad Z_0 = i$$

are collinear, since the ratio $(Z-i)/(x-i) = (r^2+1)/(x^2+1)$ is real. Hence $\mathfrak{H}_\rho$ represents the perspective with the centre i.

If $\rho > \frac{1}{2}$ the fixed points of $\mathfrak{H}_\rho$ are the two different real intersections of $\mathfrak{C}_\rho$ with the real axis. If $\rho = \frac{1}{2}$, then both fixed points coincide so that there is only the one fixed point $r = 0$. If $\rho < \frac{1}{2}$, the fixed points are

$$\left.\begin{matrix} r \\ \bar{r} \end{matrix}\right\} = \pm r = \pm i\sqrt{(1-2\rho)}.$$

They are symmetric with respect to the real axis as well as to the circle $\mathfrak{C}_\rho$.

5. Making use of the notations introduced in sections **a** and **b** show that the two ratios $(z_0-z_\infty)/(Z_0-z_\infty)$ and $(z_0-Z_\infty)/(Z_0-Z_\infty)$ are mutually reciprocal. Hence if three of the four points z_0, Z_0, z_∞, Z_∞ lie on a line, then the fourth lies on the same line.

6. Let $\mathfrak{H} = \begin{pmatrix} a & b \\ c & d \end{pmatrix}$ be the matrix of a Moebius transformation which has one single finite fixed point z_0 only. Then

$$z_0 = \frac{a-d}{2c} = \tfrac{1}{2}(z_\infty + Z_\infty).$$

7. The system of all one-dimensional projectivities in the plane is not a group. Composition of two projectivities is possible if and only if the image line of the first coincides with the line on which the second one operates. However the subsystem of all projectivities mapping a given line onto itself represents a group (cf. § 8, **d**).

8. The mapping studied in § 7, **c** appears as a *plane stereographic projection.* Let $\mathfrak{C}$ be the unit circle $z\bar{z} = 1$ and let $Z_0 = -i$. The projection ray (7.31) is then given by the equation $cz + \bar{c}\bar{z} = (\bar{c}-c)i$. Its second intersection with the unit circle is the point $\zeta = \xi + i\eta = iw$ $(w = \bar{c}/c)$. Its intersection with the real axis is

$$z = x = i\frac{\bar{c}-c}{\bar{c}+c} = i\frac{w-1}{w+1} = \frac{\zeta+\bar{\zeta}}{2-i(\zeta-\bar{\zeta})} = \frac{\xi}{1+\eta}.$$

Compare this with equation (3.31).

§ 8. Similarity and classification of Moebius transformations

a. *Introduction of a new variable.* In § 6, **e** we have investigated the manner in which the hermitian matrix of a circle changes if a new variable is introduced by means of a Moebius transformation. We shall now study the corresponding question as to how the matrix $\mathfrak{H}$ of a Moebius transformation $Z = \mathfrak{H}(z)$ changes if the variables z, Z are simultaneously substituted by new variables

$$(8.1) \qquad z^* = \mathfrak{T}(z), \qquad Z^* = \mathfrak{T}(Z), \qquad \mathfrak{T} = \begin{pmatrix} t & u \\ v & w \end{pmatrix}, \qquad |\mathfrak{T}| \neq 0.$$

Let $Z = f(z)$ be an arbitrary complex-valued function. If as a consequence of (8.1) we have the relation $Z^* = f^*(z^*)$, the function f^* will be called *similar* to f. An explicit expression of f^* can easily be found:

$$(8.11) \qquad Z^* = \mathfrak{T}(Z) = \mathfrak{T}[f(z)] = \mathfrak{T}\{f[\mathfrak{T}^{-1}(z^*)]\}.$$

If $f(z) = \mathfrak{H}(z)$ is a Moebius transformation, then any similar transformation is again a Moebius transformation $Z^* = \mathfrak{H}^*(z^*)$, whose matrix is

$$(8.12) \qquad \mathfrak{H}^* = q\mathfrak{T}\mathfrak{H}\mathfrak{T}^{-1} \qquad (q \neq 0).$$

If $\mathfrak{H} = \mathfrak{E}$ (unit matrix), then $\mathfrak{H}^* = q\mathfrak{E}$ for every $\mathfrak{T}$; thus only the identity is similar to the identity transformation.

The matrix $\mathfrak{T}\mathfrak{H}\mathfrak{T}^{-1}$ is said to be *similar* to the matrix $\mathfrak{H}$. From the elements of matrix algebra it is known (and in the present case of two-rowed matrices easily verified) that similar matrices have equal determinants and equal traces. Thus

$$\delta = |\mathfrak{H}| = ad-bc, \qquad \operatorname{tr} \mathfrak{H} = a+d = \tau$$

are similarity invariants, that is,

(8.13) $$|\mathfrak{T}\mathfrak{H}\mathfrak{T}^{-1}| = |\mathfrak{H}|, \qquad \operatorname{tr}(\mathfrak{T}\mathfrak{H}\mathfrak{T}^{-1}) = \operatorname{tr} \mathfrak{H}.$$

Two Moebius transformations with similar matrices are similar [$q = 1$ in (8.12)]. But the matrices of two similar Moebius transformations have their similarity invariants related by

$$|\mathfrak{H}^*| = q^2|\mathfrak{H}|, \qquad \operatorname{tr} \mathfrak{H}^* = q \operatorname{tr} \mathfrak{H}.$$

Thus the quotient $\dfrac{(\operatorname{tr} \mathfrak{H})^2}{|\mathfrak{H}|}$ will be a similarity invariant of the Moebius transformation with the matrix $\mathfrak{H} = \begin{pmatrix} a & b \\ c & d \end{pmatrix}$. For the identity transformation ($\mathfrak{H} = q\mathfrak{E}$) this quotient has the value.4. As the fundamental similarity invariant of the Moebius transformation we may therefore introduce the magnitude

(8.14) $$\sigma = \sigma(\mathfrak{H}) = \frac{(\operatorname{tr} \mathfrak{H})^2}{|\mathfrak{H}|} - 4 = \frac{\tau^2}{\delta} - 4 = \frac{(a-d)^2+4bc}{ad-bc},$$

so that

(8.15) $$\sigma(\mathfrak{H}^*) = \sigma(\mathfrak{H})$$

if the transformations $\mathfrak{H}$, $\mathfrak{H}^*$ are similar.

Every transformation $\mathfrak{H}$ is similar to itself (take $\mathfrak{T} = \mathfrak{E}$). If $\mathfrak{H}^*$ is similar to $\mathfrak{H}$ as expressed by (8.12), then $\mathfrak{H}$ is similar to $\mathfrak{H}^*$:

$$\mathfrak{H} = \frac{1}{q}\mathfrak{T}^{-1}\mathfrak{H}^*\mathfrak{T} = \frac{1}{q}\mathfrak{T}^{-1}\mathfrak{H}^*(\mathfrak{T}^{-1})^{-1}.$$

Finally if $\mathfrak{H}_2$ is similar to $\mathfrak{H}_1$, and $\mathfrak{H}_3$ similar to $\mathfrak{H}_2$, then $\mathfrak{H}_3$ is similar to $\mathfrak{H}_1$:
If $\mathfrak{H}_2 = \mathfrak{T}_1\mathfrak{H}_1\mathfrak{T}_1^{-1}$, $\mathfrak{H}_3 = \mathfrak{T}_2\mathfrak{H}_2\mathfrak{T}_2^{-1}$, then $\mathfrak{H}_3 = (\mathfrak{T}_2\mathfrak{T}_1)\mathfrak{H}_1(\mathfrak{T}_2\mathfrak{T}_1)^{-1}$.[3]
Thus all transformations $\mathfrak{H}$ similar to a certain $\mathfrak{H}_0$ are similar to each other; they form a *class* (that is, a class of conjugate elements) in the group of all Moebius transformations. Every element in a class can be obtained from any one element $\mathfrak{H}_0$ in the class by means of a similarity transformation with a suitable matrix $\mathfrak{T}$, that is, in the form

$$\mathfrak{H} = q\mathfrak{T}\mathfrak{H}_0\mathfrak{T}^{-1}.$$

Thus every Moebius transformation is contained in one and only one class and therefore two different classes, that is, classes having not all their elements in common, have in fact no common element at all.

[3] Briefly: similarity is reflexive, symmetric, and transitive.

b. *Normal forms of Moebius transformations.* A system of Moebius transformations which contains one and only one element out of every class, will be called a complete system of normal forms. For the determination of such a system the fixed points γ_1, γ_2 of a Moebius transformation $\mathfrak{H}$ are essential. They are the roots of the equation $\mathfrak{H}(z) = z$, that is,

$$cz^2-(a-d)z-b = 0. \tag{8.2}$$

Thus there are at most two fixed points and these may coincide. They are both finite if $c \neq 0$; one of them is infinite if $c = 0$ and $a \neq d$, and both are infinite if $c = 0$ and $a = d$. A Moebius transformation with a fixed point at infinity is integral; it is a translation if both fixed points coincide at infinity.

If z_∞ is the pole and $Z_\infty = \mathfrak{H}(\infty)$ the 'inverse pole' of $\mathfrak{H}$, then

$$\gamma_1+\gamma_2 = z_\infty+Z_\infty. \tag{8.21}$$

If γ is a fixed point of $\mathfrak{H}$, then $\mathfrak{T}(\gamma)$ is a fixed point of $\mathfrak{T}\mathfrak{H}\mathfrak{T}^{-1}$; in fact if $\mathfrak{H}(\gamma) = \gamma$, then

$$\mathfrak{T}\mathfrak{H}\mathfrak{T}^{-1}[\mathfrak{T}(\gamma)] = \mathfrak{T}\mathfrak{H}(\gamma) = \mathfrak{T}(\gamma).$$

It is also obvious that any two similar Moebius transformations have the same number of different fixed points.

Lemma. *Every Moebius transformation* $Z = \mathfrak{H}(z)$ *is similar to an integral transformation*

$$Z^* = \mathfrak{H}^*(z^*) = kz^*+t.$$

The constant k is defined by the class of $\mathfrak{H}$ *(that is, it has the same value for all similar transformations); it may be replaced by* $1/k$. *The number t is arbitrary if* $k \neq 1$.

Proof. Let $\mathfrak{H}$ be non-integral: hence $c \neq 0$. Let γ be a solution of the equation (8.2). By the transformation

$$z^* = \mathfrak{T}(z) = \frac{1}{z-\gamma}, \qquad \mathfrak{T} = \begin{pmatrix} 0 & 1 \\ 1 & -\gamma \end{pmatrix}$$

for which $\mathfrak{T}(\gamma) = \infty$, we pass over from $\mathfrak{H}$ to a similar transformation $\mathfrak{T}\mathfrak{H}\mathfrak{T}^{-1}$ which has ∞ as a fixed point and is therefore integral:

$$\mathfrak{T}\mathfrak{H}\mathfrak{T}^{-1}(z^*) = kz^*+t, \qquad \mathfrak{H}^* = \mathfrak{T}\mathfrak{H}\mathfrak{T}^{-1}.$$

Since $|\mathfrak{H}^*| = k$ and $\operatorname{tr} \mathfrak{H}^* = k+1$, we find

$$\sigma = \sigma(\mathfrak{H}) = \sigma(\mathfrak{H}^*) = \frac{(k+1)^2}{k}-4 = \frac{(k-1)^2}{k} = k+\frac{1}{k}-2. \tag{8.3}$$

For every complex number σ there is an $\mathfrak{H}$, and thus a class of $\mathfrak{T}\mathfrak{H}\mathfrak{T}^{-1}$, such that $\sigma = \sigma(\mathfrak{H})$. It is only necessary to obtain k as root of the quadratic equation

(8.3); the other root then is equal to $1/k$. Indeed $Z^* = kz^*$, matrix $\begin{pmatrix} k & 0 \\ 0 & 1 \end{pmatrix}$, and $\tilde{Z} = (1/k)\tilde{z}$ are similar transformations (cf. ex. 2). Moreover, if $k \neq 1$,

$$\begin{pmatrix} 1 & \dfrac{t}{k-1} \\ 0 & 1 \end{pmatrix} \begin{pmatrix} k & t \\ 0 & 1 \end{pmatrix} \begin{pmatrix} 1 & -\dfrac{t}{k-1} \\ 0 & 1 \end{pmatrix} = \begin{pmatrix} k & 0 \\ 0 & 1 \end{pmatrix}.$$

The constant t can thus be reduced to zero by a similarity transformation, namely the one which transforms the fixed point $\gamma_1^* = -t/(k-1)$ of $\mathfrak{H}^*$ into zero.

Now $k = 1$ if and only if $\sigma = 0$. In this case either $\mathfrak{H}^*$, and therefore $\mathfrak{H}$, is the identity transformation $\mathfrak{E}$; or $\mathfrak{H}^*$ is a translation $Z^* = z^*+t$ $(t \neq 0)$. By means of the substitution $z^* = t\tilde{z}$, $Z^* = t\tilde{Z}$ we may reduce this translation into

$$\tilde{Z} = \tilde{z}+1, \quad \text{matrix} \begin{pmatrix} 1 & 1 \\ 0 & 1 \end{pmatrix}. \tag{8.31}$$

Thus all translations are similar to each other.

Any transformation $\mathfrak{H} \neq \mathfrak{E}$, for which $\sigma = \sigma(\mathfrak{H}) = 0$, being similar to a translation (with the one fixed point ∞), must be a transformation with one single fixed point (finite or infinite). All these transformations are said to be *parabolic. All parabolic transformations form a class.* Representative of this class is the 'normal form' (8.31). There is a parabolic pencil of circles, each invariant under $\mathfrak{H}$. It is the image of the pencil of all lines parallel to the real axis by the Moebius transformation $\mathfrak{T}$ which transforms (8.31) into $\mathfrak{H}$ (cf. Figure 17).

If $\sigma \neq 0$, then $k \neq 1$ and the normal form of $\mathfrak{H}$, as well as the representative of the class of $\mathfrak{H}$, is given by

$$\tilde{Z} = k\tilde{z}, \qquad \text{matrix} \begin{pmatrix} k & 0 \\ 0 & 1 \end{pmatrix} \tag{8.32}$$

Theorem A. *Let $\mathfrak{H}_1$ and $\mathfrak{H}_2$ be two Moebius transformations, both different from the identity; they are similar to one another if and only if*

$$\sigma(\mathfrak{H}_1) = \sigma(\mathfrak{H}_2).$$

Proof. In the case $\sigma = 0$ the theorem has just been proved and by (8.15) the condition is necessary also if $\sigma \neq 0$. Suppose now that the condition is satisfied and $\sigma \neq 0$. Since

$$\sigma(\mathfrak{H}_1) = \sigma\begin{pmatrix} k_1 & 0 \\ 0 & 1 \end{pmatrix} = \sigma\begin{pmatrix} \dfrac{1}{k_1} & 0 \\ 0 & 1 \end{pmatrix}, \qquad \sigma(\mathfrak{H}_2) = \sigma\begin{pmatrix} k_2 & 0 \\ 0 & 1 \end{pmatrix}$$

$$= k_1+\frac{1}{k_1}-2, \qquad\qquad = k_2+\frac{1}{k_2}-2$$

we conclude that $k_2 = k_1$, or $k_2 = 1/k_1$. Hence $\begin{pmatrix} k_1 & 0 \\ 0 & 1 \end{pmatrix}$ and $\begin{pmatrix} k_2 & 0 \\ 0 & 1 \end{pmatrix}$, and therefore $\mathfrak{H}_1$ and $\mathfrak{H}_2$, are similar.

We shall call k the *characteristic constant* of a Moebius transformation $\mathfrak{H}$; it defines the class of $\mathfrak{H}$, but the class defines only the value of k or its reciprocal. Except for the class containing the identity $\mathfrak{E}$ only, every class defines uniquely the value of the invariant σ and is uniquely defined by the value of σ.

Another complete set of normal forms for all non-involutory Moebius transformations will be derived in § 10, ex. 14.

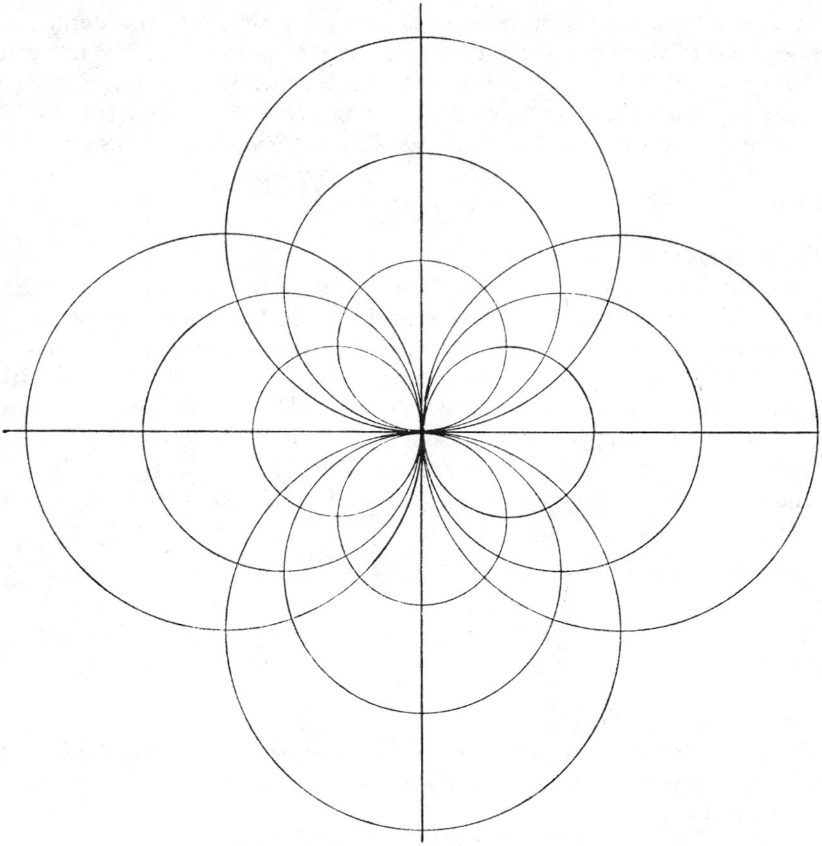

FIG. 17 Orthogonal parabolic pencils.

c. *Hyperbolic, elliptic, loxodromic transformations.* Let $\mathfrak{H}$ be a Moebius transformation with two different fixed points γ_1 and γ_2. Then $k \neq 1$, that is, $\sigma \neq 0$. It will be reduced to its similarity normal form by a transformation $\mathfrak{T}$ which shifts γ_1 into 0 and γ_2 into ∞, viz.

$$(8.4) \qquad z^* = \mathfrak{T}(z) = \frac{z-\gamma_1}{z-\gamma_2}, \qquad Z^* = \mathfrak{T}(Z).$$

Thus $\mathfrak{H}$ will be defined implicitly by

(8.41) $$\frac{Z-\gamma_1}{Z-\gamma_2} = k\frac{z-\gamma_1}{z-\gamma_2}$$

where k is the characteristic constant of $\mathfrak{H}$, or by

(8.42) $$(Z, z; \gamma_1, \gamma_2) = k \qquad \text{(cf. § 6, ex. 9).}$$

If two corresponding values, z and $Z = \mathfrak{H}(z)$ (different from γ_1, γ_2) are known, this formula gives an explicit expression for k. For instance if γ_1 and γ_2 are both finite,

(8.43) $$k = (Z_\infty, \infty; \gamma_1, \gamma_2) = \frac{Z_\infty-\gamma_1}{Z_\infty-\gamma_2} = \frac{a-c\gamma_1}{a-c\gamma_2}.$$

By substituting for γ_1, γ_2 the roots of the quadratic equation (8.2), we obtain

(8.44) $$k = \frac{\tau+\sqrt{(\sigma\delta)}}{\tau-\sqrt{(\sigma\delta)}}.$$

We shall call the transformation $\mathfrak{H}$

elliptic	if $\lvert k\rvert = 1$, that is, $-4 \leqslant \sigma < 0$	by (8.3).
proper hyperbolic	if $k > 0$, that is, $\sigma > 0$	
improper hyperbolic	if $k < 0$, that is, $\sigma \leqslant -4$	
loxodromic[4]	if $\lvert k\rvert \neq 1$ and k not real, that is, σ not real.	

Every *elliptic* transformation $\mathfrak{H}$ is similar to a rotation $Z^* = e^{i\alpha}z^*$ about 0 where $\alpha = \pm \operatorname{arc} k$, cf. § 6, **c**, 2. Thus $\mathfrak{H}$ leaves *invariant* the circles of the *hyperbolic pencil* whose point circles are at the fixed points γ_1, γ_2 of $\mathfrak{H}$. By (8.3)

(8.45) $$\sigma = -4(\sin \tfrac{1}{2}\alpha)^2.$$

A necessary and sufficient condition for $\mathfrak{H}$ to be elliptic is readily derived from (8.44):

$$|\tau-\sqrt{(\sigma\delta)}| = |\tau+\sqrt{(\sigma\delta)}|$$

which is equivalent to

(8.46) $$\tfrac{1}{2}(\bar{\tau}\sqrt{(\sigma\delta)}+\tau\overline{\sqrt{(\sigma\delta)}}) = \operatorname{Re}[\bar{\tau}\sqrt{(\sigma\delta)}] = 0$$

or $\bar{\tau}^2(\tau^2-4\delta) < 0$, or, if $\tau \neq 0$:

(8.47) $$\tau^2 \text{ and } \delta \text{ have equal arcs and } 4|\delta| > |\tau|^2.$$

[4] The 'improper hyperbolic transformations' are usually called 'loxodromic with the angle π.' By excluding them from the loxodromic transformations we may state that only those transformations are loxodromic for which there is no invariant circle (cf. below). Several other simplifications result from this arrangement which has been suggested by P. Scherk.

If $\tau = 0$, that is, $k = -1$, that is, $\sigma = -4$, the transformation will be similar to $\begin{pmatrix} -1 & 0 \\ 0 & 1 \end{pmatrix}$ which represents the transformation $Z^* = -z^*$. This is an involution. Since every transformation that is similar to an involution is itself an involution, we see that $\mathfrak{H}$ is an involution (cf. also § 6, Theorem E). Conversely every involution is similar to $\begin{pmatrix} -1 & 0 \\ 0 & 1 \end{pmatrix}$.

A *proper hyperbolic transformation* $\mathfrak{H}$ is similar to a dilatation $Z^* = \rho z^*$ $(\rho > 0)$ (cf. § 6, **c**, 3) where $\rho = k$ or $\rho = 1/k$. The pencil of circles of which each is transformed into itself by $\mathfrak{H}$ is elliptic; it consists of all circles passing through the two fixed points γ_1, γ_2 of $\mathfrak{H}$. This pencil is orthogonal to the hyperbolic pencil of invariant circles of an elliptic transformation with the same fixed points (cf. Figure 18).

An *improper hyperbolic* $\mathfrak{H}$ is similar to a transformation $Z^* = -\rho z^*$ $(\rho > 0)$ where $\rho = -k$ or $\rho = -1/k$. The two transformations

$$\mathfrak{H}_0 = \mathfrak{T}^{-1}\begin{pmatrix} \rho & 0 \\ 0 & 1 \end{pmatrix}\mathfrak{T} \qquad \text{and} \qquad \mathfrak{H} = \mathfrak{T}^{-1}\begin{pmatrix} -\rho & 0 \\ 0 & 1 \end{pmatrix}\mathfrak{T}$$

have evidently the same fixed points and the same pencil of invariant circles.

We observe that

$$\mathfrak{H}_0^{-1}\mathfrak{H} = \mathfrak{T}^{-1}\begin{pmatrix} \frac{1}{\rho} & 0 \\ 0 & 1 \end{pmatrix}\mathfrak{T}\mathfrak{T}^{-1}\begin{pmatrix} -\rho & 0 \\ 0 & 1 \end{pmatrix}\mathfrak{T} = \mathfrak{T}^{-1}\begin{pmatrix} -1 & 0 \\ 0 & 1 \end{pmatrix}\mathfrak{T} = \mathfrak{J}$$

is an involution, namely the only one in the family of all Moebius transformations that have the same fixed points γ_1, γ_2 as $\mathfrak{H}_0$ and $\mathfrak{H}$. By (8.42) $\mathfrak{J}$ is defined by

$$(Z, z; \gamma_1, \gamma_2) = -1.$$

At the same time we have shown that *for every improper hyperbolic* $\mathfrak{H}$ *there is a proper hyperbolic* $\mathfrak{H}_0$ *and an involution* $\mathfrak{J}$ *such that*

$$\mathfrak{H} = \mathfrak{H}_0\mathfrak{J}. \tag{8.48}$$

A *loxodromic transformation* $\mathfrak{H}$ possesses no invariant circle in the plane. For if

$$\mathfrak{C} = \begin{pmatrix} A & B \\ C & D \end{pmatrix}$$

is a circle in the z^*-plane, invariant under $Z^* = kz^*$, then $\mathfrak{C}$ must be equal to the circle

$$\mathfrak{C} = \begin{pmatrix} k\bar{k}A & kB \\ \bar{k}C & D \end{pmatrix}.$$

Hence either $k\bar{k} = 1$ and $B = 0$, or else $A = D = 0$ and k is real (cf. ex. 6).

Since the normal form of a loxodromic transformation is the product of a rotation $\mathfrak{R}_\alpha$ ($\alpha \neq n\pi$) and a dilatation $\mathfrak{D}_\rho$ ($\rho \neq 1$) (cf. § 6, **c**), one has

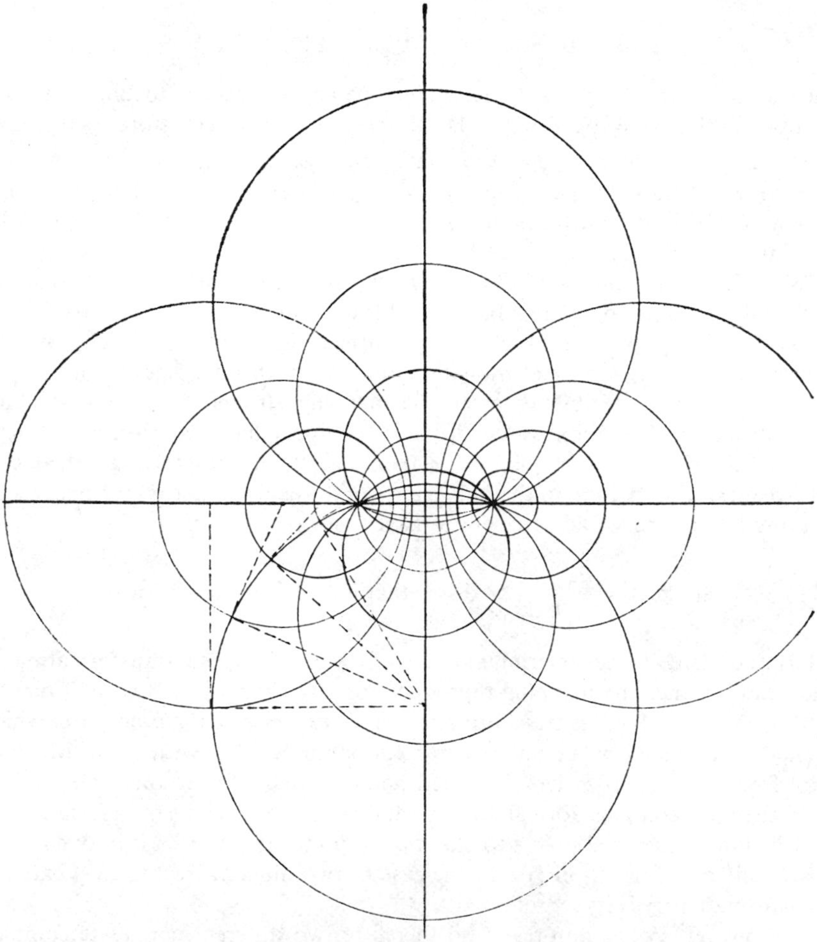

FIG. 18 Orthogonal elliptic and hyperbolic pencils.

Theorem B. *A loxodromic transformation $\mathfrak{H}$ with the fixed points γ_1, γ_2 is the product of an elliptic and a proper hyperbolic transformation, both with the same fixed points γ_1, γ_2. These two factors are commutative (cf. ex. 8).*

d. *The subgroup of the real Moebius transformations.* A Moebius transformation $Z = \mathfrak{H}(z)$ is said to be real if it has the real axis as an invariant circle.

Because a Moebius transformation preserves the orthogonality of circles, $\mathfrak{H}$ carries every pair of symmetric points z, $\bar{z}$ into symmetric points $Z = \mathfrak{H}(z)$ and $\bar{Z} = \mathfrak{H}(\bar{z})$. Thus

$$\overline{\mathfrak{H}(z)} = \mathfrak{H}(\bar{z}) \qquad \text{or} \qquad \mathfrak{H}(z) = \overline{\mathfrak{H}(\bar{z})} = \bar{\mathfrak{H}}(z).$$

Hence both matrices $\mathfrak{H}$ and $\bar{\mathfrak{H}}$ determine one and the same Moebius transformation, so that $\bar{\mathfrak{H}} = q\mathfrak{H}$ $(q \neq 0)$. This implies $\mathfrak{H} = \bar{q}\bar{\mathfrak{H}}$. Therefore $q\bar{q} = 1$ and $q = e^{i\phi}$. The matrix $\overline{e^{\frac{1}{2}i\phi}\mathfrak{H}} = e^{\frac{1}{2}i\phi}\mathfrak{H}$ which belongs to the given transformation $z = \mathfrak{H}(z)$ will thus be real. Conversely if $\mathfrak{H}$ is a real matrix, then its corresponding Moebius transformation is real. Accordingly we shall assume $\mathfrak{H}$ to be real.

With this restriction the trace of the matrix of a real Moebius transformation will be determined up to a real factor and its determinant up to a positive factor. Thus the group of real Moebius transformations contains as subgroup of index 2 the group of all real transformations with positive determinant.

There are two possibilities as to the mapping effected by a real Moebius transformation $Z = \mathfrak{H}(z)$. Either it maps the upper half plane $\operatorname{Im} z > 0$ onto itself (and the lower half plane $\operatorname{Im} z < 0$ onto itself) or it interchanges these two half planes. For $\mathfrak{H}$ to map the half planes onto themselves it is, for reasons of continuity, necessary and sufficient to have

$$\text{(8.5)} \qquad \operatorname{Im} \mathfrak{H}(i) = \frac{ad-bc}{c^2+d^2} > 0, \qquad \text{that is,} \qquad \delta = ad-bc > 0.$$

Thus the sign of the determinant $|\mathfrak{H}|$ of a real Moebius transformation $\mathfrak{H}$ indicates whether the mapping reproduces or interchanges the two half planes.

For any real Moebius transformation $\mathfrak{H}$ we expect a *real normal form* which should be obtained by a similarity transformation $\mathfrak{T}\mathfrak{H}\mathfrak{T}^{-1}$ with a *real* Moebius transformation $\mathfrak{T}$. If $\mathfrak{H}$ has real fixed points (that is, fixed points on the real axis) then the 'complex normal form' itself is real, as well as the transformation $\mathfrak{T}$ which shifts these fixed points to 0 and ∞ respectively, or to ∞ if both coincide (parabolic transformation $\mathfrak{H}$). In the case of two different real fixed points the transformation $\mathfrak{H}$ is hyperbolic.

If, however, $\mathfrak{H}$ has non-real, and therefore two different conjugate complex fixed points γ, $\bar{\gamma}$, then by (8.43) we have $|k| = 1$, and thus $\mathfrak{H}$ is *elliptic*. Hence, in virtue of (8.47), $\tau^2 < 4\delta$ and

$$\text{(8.6)} \qquad \delta = |\mathfrak{H}| > 0.$$

For every elliptic $\mathfrak{H}$ we shall now produce a similar transformation which has i and $-i$ as fixed points. This will be taken as the 'real normal form' of $\mathfrak{H}$.

Indeed let $\gamma = \gamma' + i\gamma''$ (γ' and γ'' real, $\gamma'' \neq 0$) and $\bar{\gamma} = \gamma' - i\gamma''$ be the fixed points and $k = e^{i\alpha}$ ($\alpha \neq 2n\pi$, n integral) the characteristic constant of $\mathfrak{H}$. Then a real transformation $z^* = \mathfrak{T}(z) = tz+u$ (with real t and u) can be found such that $\mathfrak{T}(\gamma) = i$. Take

$$t = \frac{1}{\gamma''}, \qquad u = -\frac{\gamma'}{\gamma''}.$$

Then the (real) transformation $\mathfrak{H}_\alpha = \mathfrak{T}\mathfrak{H}\mathfrak{T}^{-1}$ has i and $-i$ as fixed points. It will be given implicitly by (8.41), that is,

$$\text{(8.61)} \qquad \frac{Z^*-i}{Z^*+i} = e^{i\alpha}\frac{z^*-i}{z^*+i}.$$

The solution with respect to Z^* will exhibit the *real normal form*

$$\text{(8.62)} \qquad Z^* = \mathfrak{H}_\alpha(z^*), \qquad \mathfrak{H}_\alpha = \begin{pmatrix} \cos\frac{1}{2}\alpha & \sin\frac{1}{2}\alpha \\ -\sin\frac{1}{2}\alpha & \cos\frac{1}{2}\alpha \end{pmatrix}$$

of the elliptic transformation $\mathfrak{H}$. The determinant $|\mathfrak{H}_\alpha|$ has the value one.

The subgroup of all real Moebius transformations with positive determinant evidently contains the elliptic, the proper hyperbolic, and the parabolic real transformations; indeed the inequalities $\sigma > -4$ and $\delta > 0$ are equivalent. The coset of all real transformations with negative determinant consists of the improper hyperbolic Moebius transformations: $\sigma < -4$ or $\delta < 0$.

In each of the two sets there is exactly one class of involutions: **1.** *The improper hyperbolic involutions* having their fixed points on the real axis. They are all similar to the normal form $Z^* = -z^*$, whose determinant equals -1. **2.** *The elliptic involutions*, which have conjugate non-real fixed points. They are all similar to $Z^* = -1/z^*$, that is, (8.62) for $\alpha = \pi$.

Remark. The group of all real Moebius transformations is isomorphic to the group of all projectivities on the real axis (cf. § 7, ex. 7).

e. *The characteristic parallelogram.* Let $\mathfrak{H}$ be a Moebius transformation with finite fixed points γ_1, γ_2 and finite poles z_∞, Z_∞. Then by (8.21)

$$\text{(8.7)} \qquad Z_\infty - \gamma_1 = \gamma_2 - z_\infty, \qquad Z_\infty - \gamma_2 = \gamma_1 - z_\infty.$$

Thus the two pairs γ_1, γ_2 and z_∞, Z_∞ are pairs of opposite vertices of a parallelogram, the 'characteristic parallelogram' of the Moebius transformation $\mathfrak{H}$, which has already been introduced in § 7, **a**. The character of a vertex—fixed point or pole—is uniquely defined; hence every geometrical parallelogram represents two different characteristic parallelograms.

A characteristic parallelogram may degenerate into a segment and, in addition, by this process one pair of opposite vertices may coalesce into a single point of the segment (cf. § 7, **b**). Taking into account these circumstances it will be seen that not only every Moebius transformation $\mathfrak{H}$ defines uniquely its characteristic parallelogram, but also that for every given parallelogram with well defined character of the vertices there is one and only one $\mathfrak{H}$ such that the parallelogram is the characteristic one of $\mathfrak{H}$ and of $\mathfrak{H}^{-1}$, and of no other transformation.

By (8.43) the characteristic constant of $\mathfrak{H}$ is

$$k = \frac{Z_\infty - \gamma_1}{Z_\infty - \gamma_2} = \frac{z_\infty - \gamma_2}{z_\infty - \gamma_1}. \tag{8.71}$$

Therefore

$$(Z_\infty, z_\infty; \gamma_1, \gamma_2) = k^2. \tag{8.72}$$

From § 7, **b** it is known that the straight lines through γ_1 carrying a pair of adjacent sides of the parallelogram are mapped onto one another by a perspectivity from γ_2 as centre which transformation is represented by $\mathfrak{H}$.

Different types of parallelograms belong to different types of Moebius transformations.[5] If $\mathfrak{H}$ is hyperbolic [and therefore k is real ($\neq 0, 1$)], then by (8.71) the four points Z_∞, z_∞, γ_1, γ_2 are collinear. If $\mathfrak{H}$ is proper hyperbolic, $k > 0$, and z_∞, Z_∞ are outside and on different sides of the segment $\langle\gamma_1, \gamma_2\rangle$ equidistant from its midpoint. If $\mathfrak{H}$ is parabolic, then $\gamma_1 = \gamma_2$ and this point lies midway between z_∞ and Z_∞ (cf. § 7, **b**).

If $\mathfrak{H}$ is improper hyperbolic, then z_∞, Z_∞ are inner points of the segment $\langle\gamma_1, \gamma_2\rangle$ equidistant from its midpoint. For $z_\infty = Z_\infty$ the transformation $\mathfrak{H}$ will be an involution [cf. (6.52) and § 7, **b**].

If $\mathfrak{H}$ is elliptic [$|k| = 1$ ($k \neq 1, -1$)] we conclude from (8.7), (8.71) that the parallelogram must be a rhombus. It will be a square if $k = \pm i$, that is, $\sigma = -2$.

For every loxodromic Moebius transformation $\mathfrak{H}$ the characteristic parallelogram is certainly not a rhombus. It will be a rectangle if and only if $k = \pm \rho i$, that is, $\mathrm{Re}(\sigma) = -2$, [cf. (8.3)].

Since the angle of the characteristic parallelogram at the vertex z_∞ is $\theta = \operatorname{arc} k$ or $2\pi - \theta$, we conclude that similar Moebius transformations have similar characteristic parallelograms. A similarity in the z-plane is represented by a integral Moebius transformation

$$z^* = \mathfrak{T}(z) = uz + v \qquad (u \neq 0). \tag{8.73}$$

It will be shown now that the parallelogram with the vertices

$$\mathfrak{T}(\gamma_1), \qquad \mathfrak{T}(\gamma_2), \qquad \mathfrak{T}(z_\infty), \qquad \mathfrak{T}(Z_\infty) \tag{8.74}$$

will be the characteristic parallelogram of $\mathfrak{T}\mathfrak{H}\mathfrak{T}^{-1}$. Since by a suitable choice of $\mathfrak{T}$ every parallelogram in the plane, similar to γ_1, γ_2, z_∞, Z_∞, can be obtained in the form (8.74), it follows that

Theorem C. *If $\mathfrak{H}$ is a Moebius transformation with finite fixed points and finite poles, then every transformation $\mathfrak{H}_1$, similar to $\mathfrak{H}$, with finite fixed points and*

[5] Cf. E. Jacobsthal [3], II, p. 13.

finite poles can be represented in the form $\mathfrak{H}_1 = \mathfrak{T}\mathfrak{H}\mathfrak{T}^{-1}$ *where* $\mathfrak{T}$ *is a suitable integral Moebius transformation* (8.73).

Proof. The transformation $Z = \mathfrak{H}(z)$ with the fixed points γ_1, γ_2 and the pole z_∞ is implicitly given by

$$\frac{Z-\gamma_1}{Z-\gamma_2} = \frac{z_\infty-\gamma_2}{z_\infty-\gamma_1}\frac{z-\gamma_1}{z-\gamma_2} = (z, z_\infty; \gamma_1, \gamma_2) = \zeta.$$

Thus

$$Z = \frac{\gamma_2\zeta-\gamma_1}{\zeta-1} = \frac{Z_\infty z-\gamma_1\gamma_2}{z-z_\infty}$$

and

$$\mathfrak{H} = \begin{pmatrix} Z_\infty & -\gamma_1\gamma_2 \\ 1 & -z_\infty \end{pmatrix}, \qquad Z_\infty = \gamma_1+\gamma_2-z_\infty. \tag{8.75}$$

If in (8.75) we replace

$$\gamma_1 \text{ by } u\gamma_1+v, \qquad \gamma_2 \text{ by } u\gamma_2+v, \qquad z_\infty \text{ by } uz_\infty+v,$$

the matrix is changed into

$$\begin{pmatrix} uZ_\infty+v & -(u\gamma_1+v)(u\gamma_2+v) \\ 1 & -(uz_\infty+v) \end{pmatrix}$$

which indeed can be written in the form $u\mathfrak{T}\mathfrak{H}\mathfrak{T}^{-1}$ if $\mathfrak{T} = \begin{pmatrix} u & v \\ 0 & 1 \end{pmatrix}$

Now also the Moebius transformation $\mathfrak{H}^{-1}$ is similar to the transformation $\mathfrak{H}$, moreover an integral transformation $\mathfrak{T}_0$ can be found such that

$$\mathfrak{T}_0\mathfrak{H}\mathfrak{T}_0^{-1} = q\mathfrak{H}^{-1} \qquad (q \neq 0).$$

It is readily verified that the matrix

$$\mathfrak{T}_0 = \begin{pmatrix} -1 & \gamma_1+\gamma_2 \\ 0 & 1 \end{pmatrix}$$

will satisfy this condition because it interchanges the opposite vertices of the characteristic parallelogram thus preserving their character:

$$\mathfrak{T}_0(\gamma_1) = \gamma_2, \qquad \mathfrak{T}_0(\gamma_2) = \gamma_1, \qquad \mathfrak{T}_0(z_\infty) = Z_\infty.$$

Corollary. *The integral Moebius transformation* $\mathfrak{T}$ *such that*

$$\mathfrak{H}_1 = \mathfrak{T}\mathfrak{H}\mathfrak{T}^{-1}$$

is unique except when $\mathfrak{H}$ *is an involution.*

Proof. If $\mathfrak{T}_1\mathfrak{H}\mathfrak{T}_1^{-1} = q\mathfrak{T}_2\mathfrak{H}\mathfrak{T}_2^{-1}$, then

$$\mathfrak{T}_2^{-1}\mathfrak{T}_1\mathfrak{H}\mathfrak{T}_1^{-1}\mathfrak{T}_2 = \mathfrak{T}\mathfrak{H}\mathfrak{T}^{-1} = q\mathfrak{H}$$

where

$$\mathfrak{T} = \mathfrak{T}_2^{-1}\mathfrak{T}_1 = \begin{pmatrix} u & v \\ 0 & 1 \end{pmatrix}$$

and since $q^2 = 1$, u and v must satisfy the condition

$$\begin{pmatrix} u & v \\ 0 & 1 \end{pmatrix}\begin{pmatrix} a & b \\ 1 & d \end{pmatrix} = \pm\begin{pmatrix} a & b \\ 1 & d \end{pmatrix}\begin{pmatrix} u & v \\ 0 & 1 \end{pmatrix}$$

that is,

$$\begin{pmatrix} ua+v & ub+vd \\ 1 & d \end{pmatrix} = \pm\begin{pmatrix} au & av+b \\ u & v+d \end{pmatrix}.$$

Hence $u = \pm 1$ and $ua+v = \pm au$. It follows that $v = 0$ if $u = +1$; therefore $\mathfrak{T} = \mathfrak{E}$. But $v = 2a$ if $u = -1$ and this can occur only if $a+d = 0$, that is, if $\mathfrak{H}$ is an involution. Thus in this case only there are the two possibilities

$$\mathfrak{T} = \mathfrak{E} = \begin{pmatrix} 1 & 0 \\ 0 & 1 \end{pmatrix} \quad \text{and} \quad \mathfrak{T} = \begin{pmatrix} -1 & 2a \\ 0 & 1 \end{pmatrix}.$$

EXAMPLES

1. Let $\mathfrak{B}$ be any translation. Find the greatest group of Moebius transformations $\mathfrak{T}$ such that $\mathfrak{T}\mathfrak{B}\mathfrak{T}^{-1}$ is a translation for all $\mathfrak{T}$. (This group is called the 'normalizer' of the translation group within the group of all Moebius transformations.)

2. By direct calculation show that the transformation $Z = kz$ is similar to $Z^* = k_1 z^*$ if and only if $k_1 = k$ or $k_1 = 1/k$.

3. Every Moebius transformation $\mathfrak{H}$ can be written as a product of two involutions.

If $\mathfrak{H}$ is parabolic it suffices to verify the theorem for the translation $\mathfrak{B}_1 = \begin{pmatrix} 1 & 1 \\ 0 & 1 \end{pmatrix} = \mathfrak{T}\mathfrak{H}\mathfrak{T}^{-1}$. In fact $\mathfrak{B}_1 = \begin{pmatrix} -1 & 0 \\ 0 & 1 \end{pmatrix}\begin{pmatrix} -1 & -1 \\ 0 & 1 \end{pmatrix}$ and therefore

$$\mathfrak{H} = \mathfrak{T}^{-1}\begin{pmatrix} -1 & 0 \\ 0 & 1 \end{pmatrix}\mathfrak{T}\mathfrak{T}^{-1}\begin{pmatrix} -1 & -1 \\ 0 & 1 \end{pmatrix}\mathfrak{T}.$$

If $\mathfrak{H}$ is non-parabolic and its characteristic constant k is not 1 then its normal form is $\begin{pmatrix} k & 0 \\ 0 & 1 \end{pmatrix} = \begin{pmatrix} 0 & k \\ 1 & 0 \end{pmatrix}\begin{pmatrix} 0 & 1 \\ 1 & 0 \end{pmatrix}$, the product of two involutions.

4. Every improper hyperbolic real Moebius transformation is the product of a real proper hyperbolic transformation and a real involution.

5. A curve $\mathfrak{c}$ in the plane is said to be an invariant (or fixed) curve of the Moebius transformation $\mathfrak{H}$ if, for any z on $\mathfrak{c}$, $\mathfrak{H}(z)$ is also a point of $\mathfrak{c}$. This may be briefly expressed by saying that $\mathfrak{H}(\mathfrak{c}) = \mathfrak{c}$. Is it possible for $\mathfrak{H}(\mathfrak{c})$ to be only a proper part of $\mathfrak{c}$?

6. A family of invariant curves of a loxodromic Moebius transformation may be determined as follows. Let

$$z^* = r(\theta)e^{i\theta} \qquad [r(\theta) = |z^*|]$$

be the parametric representation of an invariant curve of the normal form $Z^* = kz^*$ of $\mathfrak{H}$, where θ is a real parameter. Let $k = \rho e^{i\alpha}$, $\rho = |k|$. Then

$$Z^* = \rho r(\theta) e^{i(\theta+\alpha)}$$

must also be a point of this curve. For the unknown function $r(\theta)$ we have thus the condition

$$\rho r(\theta) = r(\theta+\alpha).$$

If we choose the real constant κ so that $|k| = \rho = e^{\kappa\alpha}$, we find that

(8.8) $$r(\theta) = r_0 e^{\kappa\theta}$$

with an arbitrary positive constant, r_0, satisfies the condition.

The equation (8.8) represents a family of invariant curves in the z^*-plane in polar coordinates r, θ. These curves are *logarithmic spirals* with the fixed point 0 as centre. On returning to the z-plane by the appropriate transformation $\mathfrak{T}^{-1}$ they are mapped onto a family of invariant curves of the loxodromic transformation $\mathfrak{H}$. These 'loxodromic curves' spiral about the two fixed points of $\mathfrak{H}$.

7. Let

$$\mathfrak{H}_\rho = \begin{pmatrix} i & 2\rho-1 \\ 1 & i \end{pmatrix} \qquad \text{(cf. § 7, ex. 4).}$$

Then $\sigma(\mathfrak{H}_\rho) = \dfrac{2}{\rho} - 4$, and accordingly we find that $\mathfrak{H}_\rho$ is

improper hyperbolic	if $\rho < 0$
proper hyperbolic	if $0 < \rho < \frac{1}{2}$
parabolic	if $\rho = \frac{1}{2}$
elliptic	if $\frac{1}{2} < \rho$.

8. All Moebius transformations $\mathfrak{T}$ which commute (that is, are commutative) with a given Moebius transformation $\mathfrak{H}$, so that

(8.9) $$\mathfrak{T}\mathfrak{H} = q\mathfrak{H}\mathfrak{T} \qquad (q \neq 0),$$

form a group. By referring to the appropriate normal forms the following theorem is readily proved.

Theorem D. *If $\mathfrak{H}$ is non-parabolic and non-involutory the group of all Moebius transformations $\mathfrak{T}$ commuting with $\mathfrak{H}$ coincides with the group of all Moebius transformations having the same fixed points as $\mathfrak{H}$. If $\mathfrak{H}$ is parabolic the commuting group consists of all parabolic transformations having the same fixed point as $\mathfrak{H}$. If $\mathfrak{H}$ is an involution then the commuting group consists of all Moebius transformations $\mathfrak{T}$ having the same fixed points as $\mathfrak{H}$, and those involutions which interchange these fixed points.*

Proof. (i) In the non-parabolic case $\mathfrak{H} = \begin{pmatrix} k & 0 \\ 0 & 1 \end{pmatrix}$ $(k \neq 1)$, putting $\mathfrak{T} = \begin{pmatrix} t & u \\ v & w \end{pmatrix}$, we have (8.9) in the form

$$\begin{pmatrix} kt & u \\ kv & w \end{pmatrix} = q\begin{pmatrix} kt & ku \\ v & w \end{pmatrix}.$$

If $t \neq 0$: $\quad q = 1, \quad u = v = 0, \quad \mathfrak{T} = \begin{pmatrix} t & 0 \\ 0 & w \end{pmatrix}.$

If $t = 0$: $\quad u \neq 0, \quad v \neq 0, \quad u = qku, \quad qk = 1;$

$$kv = qv, \qquad k^2 = 1, \qquad k = -1, \quad w = 0;$$

thus

$$\mathfrak{T} = \begin{pmatrix} 0 & u \\ v & 0 \end{pmatrix}.$$

(ii) In the parabolic case $\mathfrak{H} = \begin{pmatrix} 1 & 1 \\ 0 & 1 \end{pmatrix}$, equation (8.9) becomes

$$\begin{pmatrix} t & t+u \\ v & v+w \end{pmatrix} = q\begin{pmatrix} t+v & u+w \\ v & w \end{pmatrix}.$$

Since $v \neq 0$ is impossible, we have $t \neq 0$, $v = 0$, $q = 1$, $w = t$. Hence $\mathfrak{T} = \begin{pmatrix} t & u \\ 0 & t \end{pmatrix}$, which represents a translation.

9. Let $\mathfrak{H}$ be a non-parabolic Moebius transformation and $\mathfrak{J}$ an involution. The product $\mathfrak{H}\mathfrak{J}$ is an involution if and only if the fixed points γ_1, γ_2 of $\mathfrak{H}$ form a conjugate pair of $\mathfrak{J}$.

It will be sufficient to prove the theorem for the normal form $\begin{pmatrix} k & 0 \\ 0 & 1 \end{pmatrix}$ of $\mathfrak{H}$; indeed $\begin{pmatrix} k & 0 \\ 0 & 1 \end{pmatrix}\begin{pmatrix} u & v \\ w & -u \end{pmatrix} = \begin{pmatrix} ku & kv \\ w & -u \end{pmatrix}$ is an involution if and only if $(k-1)u = 0$, that is, $u = 0$.

10. The fixed points of the Moebius transformation $\mathfrak{H} = \begin{pmatrix} a & b \\ c & d \end{pmatrix}$ are the roots of the quadratic equation (8.2), thus if $c \neq 0$:

$$(8.91) \qquad \left.\begin{matrix} \gamma_1 \\ \gamma_2 \end{matrix}\right\} = \frac{a-d}{2c} \pm \frac{1}{2c}\sqrt{[(a-d)^2+4bc]} = \frac{1}{2c}[\tau \pm \sqrt{(\delta\sigma)}] - \frac{d}{c}.$$

On the other hand, the characteristic roots of the matrix $\mathfrak{H}$ are the roots of the quadratic polynomial in λ:

$$|\lambda\mathfrak{E} - \mathfrak{H}| = \lambda^2 - \tau\lambda + \delta$$

viz.

$$\left.\begin{matrix} \lambda_1 \\ \lambda_2 \end{matrix}\right\} = \frac{\tau}{2} \pm \tfrac{1}{2}\sqrt{(\tau^2 - 4\delta)} = \tfrac{1}{2}[\tau \pm \sqrt{(\delta\sigma)}].$$

Hence

(8.92)
$$\gamma_1 = \frac{\lambda_1 - d}{c}, \qquad \gamma_2 = \frac{\lambda_2 - d}{c}.$$

Since

$$\tau = \lambda_1 + \lambda_2, \qquad \delta = \lambda_1 \lambda_2$$

it follows that

(8.93)
$$\sigma = \frac{(\lambda_1+\lambda_2)^2}{\lambda_1\lambda_2} - 4 = \frac{(\lambda_1-\lambda_2)^2}{\lambda_1\lambda_2} = \frac{\lambda_1}{\lambda_2} + \frac{\lambda_2}{\lambda_1} - 2$$

and by (8.44)

$$k = \frac{\lambda_1}{\lambda_2} \quad \left(\text{or } \frac{\lambda_2}{\lambda_1}\right) \quad \text{if} \quad \lambda_1 \neq \lambda_2.$$

If $\mathfrak{H}$ is not the identity, it is parabolic if and only if $\lambda_1 = \lambda_2$. It is an involution if and only if $\lambda_2 = -\lambda_1$.

11. A Moebius transformation $\mathfrak{H}$ having a fixed point (and therefore the poles $z_\infty = Z_\infty$) at infinity is integral and conversely. In this case the characteristic parallelogram degenerates into an angle having its vertex at the finite fixed point, but being of indefinite position otherwise. In the case of a translation the parallelogram contracts entirely into the point ∞.

12. According to § 6, ex. 14, the fixed points of an involution $Z = \mathfrak{J}(z)$ defined by two conjugate couples $z_1, Z_1; z_2, Z_2$ are the two points γ_1, γ_2 which lie in harmonic separation with each of these couples, so that

$$(z_1, Z_1; \gamma_1, \gamma_2) = -1, \qquad (z_2, Z_2; \gamma_1, \gamma_2) = -1.$$

The existence and uniqueness of the two points γ_1, γ_2 satisfying these two conditions is thus established.

Assuming that z_1, Z_1, z_2, Z_2 are not situated on the circumference of a circle F. Löbell [1] has proposed the following solution of the problem which leads to a geometrical construction of γ_1, γ_2.

If we reduce the involution $Z = \mathfrak{J}(z)$ by a Moebius transformation $z^* = \mathfrak{T}(z)$ to its normal form $Z^* = -z^*$, then

$$\mathfrak{T}(\gamma_1) = 0, \qquad \mathfrak{T}(\gamma_2) = \infty$$

are the fixed points, and any two opposite points $z^*, -z^*$ form a conjugate couple. We now choose the transforming operation $\mathfrak{T}$ so that $\mathfrak{T}(z_1) = 1$, $\mathfrak{T}(Z_1) = -1$; if $\mathfrak{T}(z_2) = \zeta$, and therefore $\mathfrak{T}(Z_2) = -\zeta$, we find the (certainly non-real) number ζ (in virtue of the invariance of the cross ratio) from the relation

$$\lambda = (z_1, Z_1; z_2, Z_2) = (1, -1; \zeta, -\zeta) = \left(\frac{1-\zeta}{1+\zeta}\right)^2,$$

to be equal to

(8.94)
$$\zeta = \frac{1+\sqrt{\lambda}}{1-\sqrt{\lambda}}.$$

Now we recognize 0 and ∞ as the intersections of two 'circles' $\mathfrak{l}_1$, $\mathfrak{l}_2$, where $\mathfrak{l}_1$ is the real axis, that is, the circle of the elliptic pencil through 1 and -1 which bisects the angle between those two circles of this pencil that pass through ζ

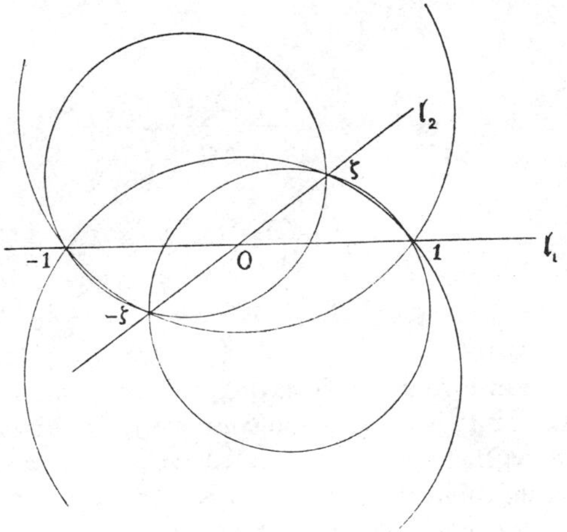

FIG. 19

and $-\zeta$ respectively; and $\mathfrak{l}_2$ is the line belonging to the elliptic pencil through

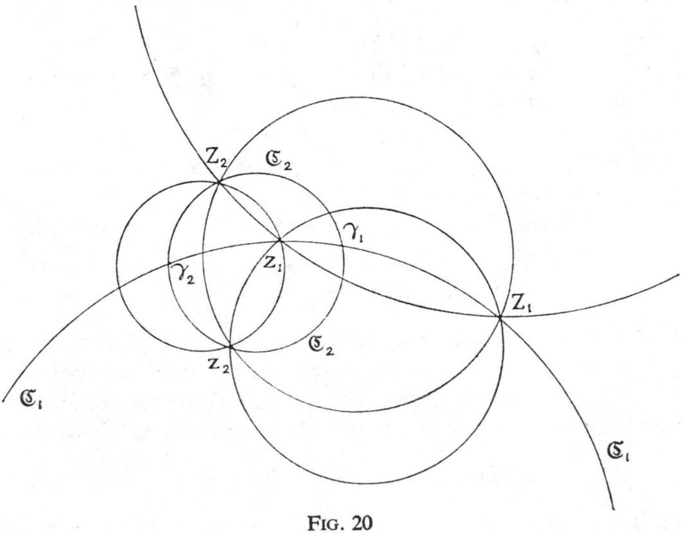

FIG. 20

ζ and $-\zeta$, that is, the circle that bisects the angle between those two circles of the pencil which pass through 1 and -1 respectively (cf. Figure 19).

By inverting the transformation $\mathfrak{T}$ we arrive at the following situation in the z-plane (Figure 20). The points $\gamma_1 = \mathfrak{T}^{-1}(0)$ and $\gamma_2 = \mathfrak{T}^{-1}(\infty)$ are the intersections of the two circles $\mathfrak{C}_1 = \mathfrak{T}^{-1}(\mathfrak{l}_1)$, $\mathfrak{C}_2 = \mathfrak{T}^{-1}(\mathfrak{l}_2)$ which can be determined in the following way. $\mathfrak{C}_1$ is one of the circles of the elliptic pencil through z_1, Z_1 that bisects the angle between those two circles of this pencil which pass through z_2, Z_2 respectively. Further, $\mathfrak{C}_2$ is a circle of the elliptic pencil through z_2, Z_2 that bisects the angle between those two circles of this pencil which pass through z_1, Z_1 respectively. The choice of the two bisecting circles (among the two for $\mathfrak{C}_1$ and the two for $\mathfrak{C}_2$, cf. § 4, ex. 8) is unique, because γ_1, γ_2 must separate z_1, Z_1 on $\mathfrak{C}_1$ and z_2, Z_2 on $\mathfrak{C}_2$ in order to have the cross ratios negative, viz. -1 (cf. § 6, ex. 15).

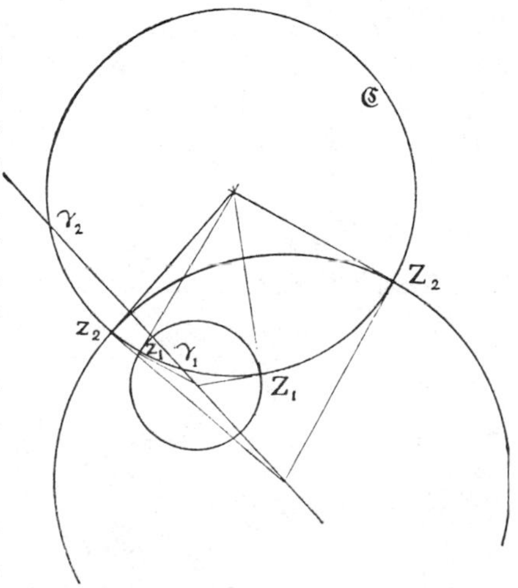

FIG. 21

If the two couples $z_1, Z_1; z_2, Z_2$ lie on a circle $\mathfrak{C}$ then there are two possibilities.

1. Inclusion: $\lambda = (z_1, Z_1; z_2, Z_2) > 0$. In this case ζ is real and 0 and ∞ are the point circles of the hyperbolic pencil of all concentric circles about 0, which are orthogonal to the real axis. Thus γ_1, γ_2 are the point circles of the hyperbolic pencil orthogonal to $\mathfrak{C}$ containing the circles through z_1, Z_1 and z_2, Z_2. The points γ_1, γ_2 are thus the intersections of $\mathfrak{C}$ with the line joining the centres of those two orthogonal circles (cf. Figure 21).

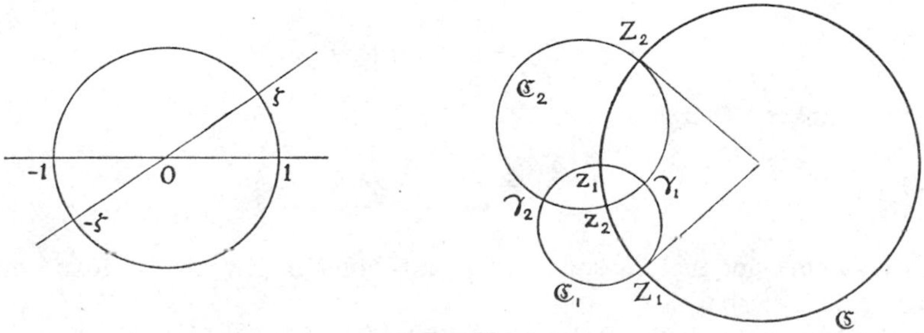

FIG. 22

2. Separation: $\lambda < 0$. The four points $-1, 1; -\zeta, \zeta$ separate each other on the unit circle. Then γ_1, γ_2 are the intersections of the two circles $\mathfrak{C}_1$, $\mathfrak{C}_2$ that cut $\mathfrak{C}$ orthogonally at z_1, Z_1 and z_2, Z_2 respectively (cf. Figure 22).

13. Let $\mathfrak{H}$ be a Moebius transformation with finite pole z_∞ ($c \neq 0$). Then the pole z'_∞ of $\mathfrak{H}^{-1}$ is finite. If $\mathfrak{H}$ is not an involution ($z_\infty \neq z'_\infty$), then the straight line through z_∞ and z'_∞ is invariant by $\mathfrak{H}$ if and only if $\mathfrak{H}$ is not loxodromic.

§ 9. Classification of anti-homographies

a. *Anti-homographies.* The definition and elementary mapping properties of the anti-homographies have been stated and discussed in § 6 (in particular § 6, **a**, Remark 2; § 6, **d**; and § 6, ex. 12).

If $z_1 = \mathfrak{H}_1(\bar{z})$ and $Z = \mathfrak{H}_2(\bar{z}_1)$ are two anti-homographies, then their product

$$Z = \mathfrak{H}_2(\overline{\mathfrak{H}_1(\bar{z})}) = \mathfrak{H}_2\overline{\mathfrak{H}}_1(z) = \mathfrak{H}_3(z)$$

is a Moebius transformation with the matrix

$$\mathfrak{H}_3 = q\mathfrak{H}_2\overline{\mathfrak{H}}_1 \qquad (q \neq 0). \tag{9.1}$$

In particular, if $\mathfrak{H}_1 = \mathfrak{H}_2 = \mathfrak{H} = \begin{pmatrix} a & b \\ c & d \end{pmatrix}$ we obtain the Moebius transformation

$$Z = \mathfrak{H}\overline{\mathfrak{H}}(z)$$

which we shall call 'the square' of the anti-homography $Z = \mathfrak{H}(\bar{z})$. Its determinant is

$$\delta_2 = |\mathfrak{H}\overline{\mathfrak{H}}| = |\mathfrak{H}||\overline{\mathfrak{H}}| = |\mathfrak{H}|\overline{|\mathfrak{H}|} = |\delta|^2 > 0 \tag{9.11}$$

and its trace

$$\tau_2 = \operatorname{tr}(\mathfrak{H}\overline{\mathfrak{H}}) = a\bar{a}+d\bar{d}+b\bar{c}+\bar{b}c \tag{9.12}$$

is a real number. Therefore

$$\sigma(\mathfrak{H}\overline{\mathfrak{H}}) = \frac{\tau_2^2}{\delta_2}-4 \geqslant -4 \tag{9.13}$$

whence we conclude that the square of an anti-homography is never loxodromic or improper hyperbolic.

By the Moebius transformation $z^* = \mathfrak{T}(z)$, $Z^* = \mathfrak{T}(Z)$ we introduce the new variables z^*, Z^*. If z and Z are connected by the anti-homography $Z = \mathfrak{H}(\bar{z})$, then the anti-homography

$$Z^* = \mathfrak{T}(\mathfrak{H}(\bar{z})) = \mathfrak{T}\mathfrak{H}(\overline{\mathfrak{T}^{-1}(z^*)}) = \mathfrak{T}\mathfrak{H}\overline{\mathfrak{T}}^{-1}(\overline{z^*}) = \mathfrak{H}^*(\overline{z^*}) \tag{9.2}$$

is said to be similar to the anti-homography $Z = \mathfrak{H}(\bar{z})$. The matrix

(9.21) $$\mathfrak{H}^* = \mathfrak{T}\mathfrak{H}\mathfrak{T}^{-1}$$

is said to be related[6] to the matrix $\mathfrak{H}$. By (9.21)

$$\mathfrak{H}^*\overline{\mathfrak{H}^*} = \mathfrak{T}\mathfrak{H}\mathfrak{T}^{-1}\mathfrak{T}\overline{\mathfrak{H}}\mathfrak{T}^{-1} = \mathfrak{T}\mathfrak{H}\overline{\mathfrak{H}}\mathfrak{T}^{-1};$$

thus to similar anti-homographies, correspond similar squares.

Similarity of anti-homographies and relatedness of matrices are reflexive, symmetric, and transitive (cf. §8, **a**, p. 61). Hence there is a division of the system of all anti-homographies into classes of similar anti-homographies so that different classes contain no common element. Each class can be characterized by one of its elements. A complete set of normal forms will be established.

b. *Anti-involutions.* By definition an anti-involution is an involutory anti-homography, that is, a transformation $Z = \mathfrak{K}(\bar{z})$, $|\mathfrak{K}| = \delta \neq 0$, whose square is the identity:

(9.3) $$\mathfrak{K}\overline{\mathfrak{K}} = \mu\mathfrak{E}.$$

The 'multiplier' μ is a real number for which $\delta\bar{\delta} = \mu^2$. A two-rowed complex matrix $\mathfrak{K}$ that satisfies the condition (9.3) is called a *circle matrix.*[7]

If a circle is given by the hermitian matrix

$$\mathfrak{C} = \begin{pmatrix} A & B \\ C & D \end{pmatrix}, \qquad |\mathfrak{C}| = \Delta \neq 0,$$

it determines the inversion (2.21) with the matrix

(9.31) $$\mathfrak{K} = \begin{pmatrix} -C & -D \\ A & B \end{pmatrix}, \qquad |\mathfrak{K}| = |\mathfrak{C}|,$$

that is, the transformation $Z = \mathfrak{K}(\bar{z})$, which is an anti-involution (cf. § 2, **a**):

$$\mathfrak{K}\overline{\mathfrak{K}} = -|\mathfrak{C}|\mathfrak{E}.$$

Thus $\mathfrak{K}$ is a circle matrix with the multiplier

(9.32) $$\mu = -|\mathfrak{C}|$$

which is positive if the fundamental circle $\mathfrak{C}$ of the inversion $Z = \mathfrak{K}(\bar{z})$ is real, negative if it is imaginary. The fundamental circle is the set of all fixed points of the inversion. Its equation can therefore be written in the form

9.33) $$\mathfrak{K}(\bar{z}) = z.$$

[6] 'Verwandt,' cf. E. Jacobsthal [2].

[7] 'Kreismatrix,' cf. Jacobsthal [2].

Every hermitian matrix $\mathfrak{C}$ with non-vanishing determinant is turned into a circle matrix $\mathfrak{K}$ by inverting the sign of one row and interchanging the two rows. The proof of the following theorem shows that with a slight modification this process may be reversed.

Theorem A. *Every anti-involution $Z = \mathfrak{K}(\bar{z})$ represents an inversion.*

Proof. By hypothesis $\mathfrak{K}$ is a circle matrix. Let $\mathfrak{K} = \begin{pmatrix} a & b \\ c & d \end{pmatrix}$. First we assume that $c = |c|e^{i\phi}$ is non-zero. Let $\mathfrak{K}_0 = \begin{pmatrix} a_0 & b_0 \\ c_0 & d_0 \end{pmatrix} = e^{-i\phi}\,\mathfrak{K}$ so that $c_0 = |c| > 0$. The matrix $\mathfrak{K}_0$ represents the same anti-involution as $\mathfrak{K}$ and by (9.3) $\mathfrak{K}_0\bar{\mathfrak{K}}_0 = \mu\mathfrak{E}$, that is,

$$a_0\bar{a}_0 + b_0c_0 = \mu, \qquad a_0\bar{b}_0 + b_0\bar{d}_0 = 0$$
$$c_0\bar{a}_0 + d_0c_0 = 0, \qquad c_0\bar{b}_0 + d_0\bar{d}_0 = \mu.$$

Since μ is real, it follows that $b_0 = \dfrac{1}{c_0}(\mu - a\bar{a})$ is real. Moreover $c_0(\bar{a}_0 + d_0) = 0$; hence $\bar{a}_0 = -d_0$. Thus the matrix

$$\mathfrak{C} = \begin{pmatrix} c_0 & d_0 \\ -a_0 & -b_0 \end{pmatrix}$$

will be hermitian. The given anti-involution will appear as the inversion with respect to the circle $\mathfrak{C}$.

If $c = 0$ and $b \neq 0$, let $b = |b|e^{i\phi}$. Let $\mathfrak{K}_0 = e^{-i\phi}\,\mathfrak{K}$. Then $b_0 = |b| > 0$ and the previous argument can be repeated.

If $b = 0$ and $c = 0$, it follows that $|a| = |d|$. Let $a = |a|e^{i\alpha}$, $d = |a|e^{i\beta}$. Then it is found that the matrix $\mathfrak{C} = ie^{-\frac{1}{2}i(\alpha+\beta)}\begin{pmatrix} 0 & d \\ -a & 0 \end{pmatrix}$ is hermitian.

Theorem B. *The matrix*

$$\mathfrak{K}^* = \mathfrak{T}\mathfrak{K}\mathfrak{T}^{-1},$$

related to a circle matrix $\mathfrak{K}$, is itself a circle matrix with the same multiplier as $\mathfrak{K}$. If $\mathfrak{K}$ corresponds to the hermitian matrix $\mathfrak{C}$, then $\mathfrak{K}^*$ corresponds to the hermitian matrix

$$\mathfrak{C}^* = \mathfrak{S}'\mathfrak{C}\bar{\mathfrak{S}}, \qquad \mathfrak{S} = \mathfrak{T}^{-1} \qquad \text{(cf. § 6, e.)}$$

Proof. By (9.3) $\mathfrak{K}^*\overline{\mathfrak{K}^*} = \mathfrak{T}\mathfrak{K}\bar{\mathfrak{K}}\mathfrak{T}^{-1} = \mu\mathfrak{E}$. Further, the equation of the circle with the hermitian matrix $\mathfrak{C}$ may be written in the form (9.33). Under the transformation $z^* = \mathfrak{T}(z)$ it becomes

$$z^* = \mathfrak{T}(\mathfrak{K}(\bar{z})) = \mathfrak{T}(\mathfrak{K}(\overline{\mathfrak{T}^{-1}(z^*)})) = \mathfrak{T}\mathfrak{K}\mathfrak{T}^{-1}(\overline{z^*}) = \mathfrak{K}^*(\overline{z^*}).$$

On the other hand, by (6.42), the hermitian matrix of this circle is the matrix $\mathfrak{C}^*$.

c. *Normal forms of non-involutory anti-homographies.* As in the case of the Moebius transformation, so also for the classification of the anti-homographies the fixed points are fundamental. Let $Z = \mathfrak{H}(\bar{z})$ be a non-involutory antihomography and γ_1 be a fixed point of its square $Z = \mathfrak{H}\bar{\mathfrak{H}}(z)$. Let $\mathfrak{H}(\bar{\gamma}_1) = \gamma_2$ and thus $\mathfrak{H}(\bar{\gamma}_2) = \gamma_1$. Then either $\gamma_2 = \gamma_1$ and γ_1 is a fixed point of $Z = \mathfrak{H}(\bar{z})$; or $\gamma_2 \neq \gamma_1$, in which case γ_1 and γ_2 form a 'conjugate pair' of $Z = \mathfrak{H}(\bar{z})$, that is, they are interchanged by the anti-homography.

Conversely: Every fixed point, and the two points of a conjugate pair of $Z = \mathfrak{H}(\bar{z})$ are fixed points of the square $Z = \mathfrak{H}\bar{\mathfrak{H}}(z)$. From § 9, **a** it is known that this Moebius transformation is either

$$\text{(i) proper hyperbolic: } \mathfrak{T}\mathfrak{H}\bar{\mathfrak{H}}\mathfrak{T}^{-1} = \begin{pmatrix} \rho_2 & 0 \\ 0 & 1 \end{pmatrix} \qquad (\rho_2 > 0)$$

or

$$\text{(ii) parabolic} \qquad \mathfrak{T}\mathfrak{H}\bar{\mathfrak{H}}\mathfrak{T}^{-1} = \begin{pmatrix} 1 & 1 \\ 0 & 1 \end{pmatrix}$$

or

$$\text{(iii) elliptic} \qquad \mathfrak{T}\mathfrak{H}\bar{\mathfrak{H}}\mathfrak{T}^{-1} = \begin{pmatrix} e^{i\beta} & 0 \\ 0 & 1 \end{pmatrix} \qquad (\beta \neq 2n\pi).$$

From the statements about the fixed points of a non-involutory anti-homography $Z = \mathfrak{H}(\bar{z})$ we conclude that there are the following three possibilities: it has either

(i) exactly two different fixed points and no conjugate pair: hyperbolic anti-homography;

or

(ii) exactly one fixed point and no conjugate pair: parabolic anti-homography;

or

(iii) no fixed point and exactly one conjugate pair: elliptic anti-homography.

We shall see now that this division of the anti-homographies corresponds to the division of their squares as given above.

(i) If γ_1, γ_2 are the two fixed points of the given *hyperbolic anti-homography* $Z = \mathfrak{H}(\bar{z})$, we may apply the transformation $z_1 = \mathfrak{T}_1(z) = (z-\gamma_1)/(z-\gamma_2)$, $Z_1 = \mathfrak{T}_1(Z)$ which reduces the transformation into

$$Z_1 = \mathfrak{T}_1\mathfrak{H}\bar{\mathfrak{T}}_1^{-1}(\bar{z}_1) = k\bar{z}_1, \qquad k = |k|e^{i\alpha},$$

that has 0 and ∞ as fixed points (cf. § 8, **c**). If $\alpha \neq 2n\pi$, the transformation $z^* = e^{-i(\alpha/2)}z_1$, $Z^* = e^{-i(\alpha/2)}Z_1$ will lead us to

$$Z^* = |k|e^{i(\alpha/2)}\bar{z}_1 = |k|\overline{z^*} = \mathfrak{H}^*(\overline{z^*}) \tag{9.4}$$

which is the normal form. The positive constant $|k| = \rho$ may be replaced by $1/\rho$, as in § 8, **b**.

The square of the normal form $Z^* = \mathfrak{H}^*(\overline{z^*})$ is evidently the proper hyperbolic normal form of $\mathfrak{H}\bar{\mathfrak{H}}$, that is, the dilatation $\mathfrak{H}^*\bar{\mathfrak{H}}^* = \begin{pmatrix} \rho_2 & 0 \\ 0 & 1 \end{pmatrix}$ and $\rho_2 = \rho^2$.
The similarity class of $\mathfrak{H}(\bar{z})$ is thus uniquely defined by the similarity class of the Moebius transformation $\mathfrak{H}\bar{\mathfrak{H}}$, that is, by the value $\sigma(\mathfrak{H}\bar{\mathfrak{H}}) > 0$.

(ii) The anti-homography $Z = \mathfrak{H}(\bar{z})$ and its square are *parabolic* at the same time, and for both the normal form is obtained by shifting the only fixed point into infinity. Thus $\mathfrak{H}^*$ has to satisfy the condition

$$\mathfrak{H}^*\overline{\mathfrak{H}^*} = \begin{pmatrix} 1 & 1 \\ 0 & 1 \end{pmatrix}.$$

To have an infinite fixed point, $\mathfrak{H}^*(\overline{z^*})$ must be of the form $a\,\overline{z^*} + b$. Thus

$$\mathfrak{H}^*\overline{\mathfrak{H}^*} = \begin{pmatrix} a\bar{a} & a\bar{b}+b \\ 0 & 1 \end{pmatrix} = \begin{pmatrix} 1 & 1 \\ 0 & 1 \end{pmatrix}$$

and therefore

$$a = \frac{1-b}{\bar{b}}, \qquad a\bar{a} = \frac{1-b-\bar{b}+b\bar{b}}{b\bar{b}} = 1$$

or

$$(9.41) \qquad b+\bar{b} = 1, \qquad a = 1, \qquad \mathfrak{H}_1^* = \begin{pmatrix} 1 & b \\ 0 & 1 \end{pmatrix}.$$

By means of the transformation $\mathfrak{T} = \begin{pmatrix} 1 & -b/2 \\ 0 & 1 \end{pmatrix}$ the matrix is reduced into

$$(9.42) \qquad \mathfrak{H}^* = \begin{pmatrix} 1 & \frac{1}{2} \\ 0 & 1 \end{pmatrix}, \qquad \text{and} \qquad Z^* = \overline{z^*}+\tfrac{1}{2}$$

is the parabolic normal form. The parabolic class of anti-homographies $Z = \mathfrak{H}(\bar{z})$ is characterized by $\sigma(\mathfrak{H}\overline{\mathfrak{H}}) = 0$.

(iii) Finally, if $\mathfrak{H}\overline{\mathfrak{H}}$ is *elliptic* with $\mathfrak{H}^*\overline{\mathfrak{H}^*} = \begin{pmatrix} e^{i\beta} & 0 \\ 0 & 1 \end{pmatrix}$ as its normal form, then the normal form $Z^* = \mathfrak{H}^*(\overline{z^*})$ has 0 and ∞ (the fixed points of $\mathfrak{H}^*\overline{\mathfrak{H}^*}$) as its conjugate pair, whence

$$\mathfrak{H}^*(\overline{z^*}) = \frac{\kappa}{\overline{z^*}} \qquad (\kappa \neq 1).$$

Thus

$$\mathfrak{H}^* = \begin{pmatrix} 0 & \kappa \\ 1 & 0 \end{pmatrix}, \qquad \mathfrak{H}\overline{\mathfrak{H}^*} = \begin{pmatrix} \kappa & 0 \\ 0 & \bar{\kappa} \end{pmatrix} = \bar{\kappa}\begin{pmatrix} \frac{\kappa}{\bar{\kappa}} & 0 \\ 0 & 1 \end{pmatrix}$$

or

$$\frac{\kappa}{\bar{\kappa}} = e^{i\beta}, \qquad \kappa = |\kappa|e^{i(\beta/2)}.$$

By means of the additional transformation $\mathfrak{T}_1 = \begin{pmatrix} |\kappa|^{-\frac{1}{2}} & 0 \\ 0 & 1 \end{pmatrix}$ the modulus $|\kappa|$ may be reduced to unity; thus

$$(9.43) \qquad \mathfrak{T}_1\mathfrak{H}^*\mathfrak{T}_1^{-1} = \sqrt{|\kappa|}\begin{pmatrix} 0 & e^{i(\beta/2)} \\ 1 & 0 \end{pmatrix}, \qquad \text{and} \qquad Z^* = \frac{e^{i(\beta/2)}}{\overline{z^*}}$$

is the elliptic normal form. The elliptic classes are characterized by the values $\sigma(\mathfrak{H}\bar{\mathfrak{H}})$ in the interval $-4 \leqslant \sigma < 0$.

As a result of the preceding discussion the following theorem may be stated.

Theorem C. *Two non-involutory anti-homographies $Z = \mathfrak{H}_1(\bar{z})$ and $Z = \mathfrak{H}_2(\bar{z})$ are similar if and only if their squares are similar Moebius transformations.*

d. *Normal forms of circle matrices and anti-involutions.* The first statement of Theorem B may be inverted.

Theorem D. *If two circle matrices have the same multiplier, then they are related.*

Proof. Let $\mathfrak{K}$ be a circle matrix and μ its multiplier. If $\mu > 0$, we observe that $\sqrt{(\mu)}\mathfrak{E}$ is a circle matrix with μ as multiplier. We show that it is related to $\mathfrak{K}$. Indeed, put $\mathfrak{T}^{-1} = \mathfrak{S} = \mathfrak{K}+\sqrt{(\mu)}\mathfrak{E}$; then by (9.3)

$$\mathfrak{K}\bar{\mathfrak{S}} = \mathfrak{K}\bar{\mathfrak{K}}+\sqrt{(\mu)}\mathfrak{K} = \mu\mathfrak{E}+\sqrt{(\mu)}\mathfrak{K} = \sqrt{(\mu)}\mathfrak{S}$$

or $\mathfrak{T}\mathfrak{K}\bar{\mathfrak{T}}^{-1} = \sqrt{(\mu)}\mathfrak{E}$.

If $\mu = -\rho^2 < 0$, $\rho > 0$, $\mathfrak{J} = \begin{pmatrix} 0 & 1 \\ -1 & 0 \end{pmatrix}$, we observe that the matrix $\rho i\mathfrak{J}$ is a circle matrix with the multiplier μ; we show that $\mathfrak{K}$ is related to this matrix. Let $\mathfrak{S} = \mathfrak{K}\mathfrak{J}+\rho i\mathfrak{E}$; since $\mathfrak{J}^2 = -\mathfrak{E}$, it follows that

$$\mathfrak{K}\bar{\mathfrak{S}} = \mu\mathfrak{J}-\rho i\mathfrak{K} = \rho i(\rho i\mathfrak{E}+\mathfrak{K}\mathfrak{J})\mathfrak{J} = \rho i\mathfrak{S}\mathfrak{J}.$$

Thus the theorem is proved.

We conclude that there are two similarity classes of anti-involutions (inversions) $Z = \mathfrak{K}(\bar{z})$:

(i) The inversions with a real fundamental circle ($\mu > 0$). The normal form is $Z^* = \overline{z^*}$, the corresponding fundamental circle is the real axis $z^* = \overline{z^*}$ ($\mathfrak{K}^* = \sqrt{(\mu)}\mathfrak{E}$).

(ii) The inversions with imaginary fundamental circle ($\mu < 0$). The normal form is $Z^* = -1/\overline{z^*}$ ($\mathfrak{K}^* = \sqrt{(\mu)}\mathfrak{J}$). The corresponding circle is the imaginary unit circle $z^*\overline{z^*}+1 = 0$. Cf. § 6, e.

e. *Moebius transformations and anti-homographies as products of inversions.* In § 6, **c** it has been shown that each of the simple type Moebius transformations can be written as a product of two inversions. To the simple types considered in § 6, **c** we may add now the 'improper dilatation'

$$Z^* = -\rho z^* \qquad (\rho > 0),$$

that is, the normal form of an improper hyperbolic Moebius transformation, which can also be represented as a product of two inversions. As first fundamental circle we choose the imaginary unit circle about 0, yielding the inversion

$$z_1^* = -\frac{1}{\overline{z^*}}.$$

For the second inversion we take the circle with radius $\sqrt{\rho} > 0$ about 0; then indeed

$$Z^* = \frac{\rho}{\overline{z_1^*}} = -\rho z^*.$$

Thus if a given Moebius transformation $\mathfrak{H}$ is non-loxodromic, then there is a Moebius transformation $\mathfrak{T}$ such that

$$\mathfrak{T}\mathfrak{H}\mathfrak{T}^{-1}(z^*) = \mathfrak{H}^*(z^*) = \begin{cases} z^*+1 & \text{if } \mathfrak{H} \text{ is parabolic} \\ kz^* & \text{where } k \text{ is real or } |k| = 1, \quad k \neq 1. \end{cases}$$

Now

$$\mathfrak{H}^* = \mathfrak{J}_2^*\overline{\mathfrak{J}_1^*}$$

where $\mathfrak{J}_1$, $\mathfrak{J}_2$ are two inversions; hence

$$\mathfrak{H} = \mathfrak{T}^{-1}\mathfrak{H}^*\mathfrak{T} = (\mathfrak{T}^{-1}\mathfrak{J}^*\overline{\mathfrak{T}})(\overline{\mathfrak{T}}^{-1}\overline{\mathfrak{J}_1^*}\mathfrak{T})$$

is also a product of two inversions.

If the fundamental circle of an inversion belongs to a pencil, then we shall say that the inversion is an *inversion within this pencil.* The inversions of which the elementary types are products, are inversions: within the pencil of all straight lines through, or circles about, 0 in the case of a rotation, or dilatation; within the pencil of all lines parallel to the imaginary axis in the case of a translation. By the transformation $\mathfrak{T}^{-1}$ this pencil is transformed into the elliptic (hyperbolic, parabolic) pencil of circles which are interchanged by the given elliptic (hyperbolic, parabolic) Moebius transformation $\mathfrak{H}$. The pencil is defined by the fixed points γ_1, γ_2 of $\mathfrak{H}$. In the elliptic case it is the pencil of all circles through γ_1 and γ_2; in the hyperbolic case the pencil of which γ_1, γ_2 are the point circles.

Theorem E. *Every elliptic* (*hyperbolic, parabolic*) *Moebius transformation* $Z = \mathfrak{H}(z)$ *can be represented as the product of two inversions within the elliptic* (*hyperbolic, parabolic*) *pencil of circles which is orthogonal to the hyperbolic* (*elliptic, parabolic*) *pencil of circles, invariant under the transformation* $\mathfrak{H}$. (*Cf.* § 8, **b**, *Figure* 17*;* **c**, *Figure* 18.)

The converse of this theorem is also true:

Theorem F. *The product of any two different inversions is a non-loxodromic Moebius transformation.*

Proof. Let $\mathfrak{C}_1$, $\mathfrak{C}_2$ be the fundamental circles of the two inversions. Their product is then a Moebius transformation $\mathfrak{H}$ within the pencil generated by $\mathfrak{C}_1$ and $\mathfrak{C}_2$, which means that this is the pencil of circles interchanged by the transformation $\mathfrak{H}$. By turning to the normal form of the pencil (that is, all lines through, or all circles about, 0, or all lines parallel to the imaginary axis) the Moebius transformation $\mathfrak{H}$ is reduced into one of the simple types; hence it is non-loxodromic.

The normal form of a loxodromic transformation is the (commutative) product of a rotation and a proper dilatation. As a consequence of the preceding two theorems we have the following:

Corollary. *Every loxodromic Moebius transformation is the product of four, and never less than four, inversions.*

Now let $Z = \mathfrak{H}(\bar{z})$ be a non-involutory anti-homography. According to § 9, **c**, with a suitable Moebius transformation $\mathfrak{T}$:

$$\mathfrak{T}\mathfrak{H}\mathfrak{T}^{-1}(\overline{z^*}) = \begin{cases} \rho\overline{z^*} \quad (\rho > 0) & \text{in the hyperbolic case} \\ \overline{z^*} + \frac{1}{2} & \text{in the parabolic case} \\ \dfrac{e^{i(\beta/2)}}{\overline{z^*}} & \text{in the elliptic case.} \end{cases}$$

Each of these normal forms is readily seen to be the product of exactly three inversions. The same is therefore true for the given anti-homography. Thus

Theorem G. *Every non-involutory anti-homography is the product of exactly three inversions.*

f. *The groups of a pencil.* It is readily seen that the system of all Moebius transformations $\mathfrak{H}$ with given fixed points γ_1, γ_2 (or fixed point γ) is a group. Indeed if $\mathfrak{H}_1(\gamma) = \gamma$ and $\mathfrak{H}_2(\gamma) = \gamma$ then $\mathfrak{H}_2\mathfrak{H}_1(\gamma) = \gamma$. Besides $\mathfrak{H}$ the system contains the inverse $\mathfrak{H}^{-1}$, and therefore the identity $\mathfrak{E}$.

Let us denote by $\mathscr{H}(\gamma_1, \gamma_2) = \mathscr{H}$ the group of all Moebius transformations having γ_1, γ_2 as fixed points, and $\gamma_1 \neq \gamma_2$. The elements of this group are elliptic or hyperbolic or loxodromic. By one and the same transformation $\mathfrak{T}$, viz. (8.4), they are all reduced to their normal forms $Z^* = kz^*$. Thus the group $\mathscr{H}$, being similar, and therefore isomorphic to the group of these normal forms, is commutative. It appears as the direct product of the group $\mathscr{H}_e = \mathscr{H}_e(\gamma_1, \gamma_2)$ of the elliptic, and the group $\mathscr{H}_h^+ = \mathscr{H}_h^+(\gamma_1, \gamma_2)$ of the proper hyperbolic transformations with γ_1, γ_2 as fixed points.

All hyperbolic transformations with the fixed points γ_1, γ_2 form a group $\mathscr{H}_h = \mathscr{H}_h(\gamma_1, \gamma_2)$ of which $\mathscr{H}_h^+$ is a subgroup of index 2. Its coset $\mathscr{H}_h^-$ contains all improper hyperbolic transformations. By (8.48)

$$\mathscr{H}_h^- = \mathscr{H}_h^+\mathfrak{J},$$

where $\mathfrak{J}$ is an involution.

Every element of $\mathscr{H}_e$ is product of two inversions within the elliptic pencil of all circles through γ_1, γ_2. Every element of $\mathscr{H}_h$ can be written as the product of two inversions within the hyperbolic pencil which is orthogonal to the elliptic pencil of circles through γ_1, γ_2. The group $\mathscr{H}_h^+$ consists of all proper hyperbolic transformations which appear as the product of two inversions within the same hyperbolic pencil, but with real fundamental circles.

If $\gamma_1 = \gamma_2 = \gamma$, we obtain the group $\mathscr{H}_p = \mathscr{H}_p(\gamma)$ of parabolic transformations with γ as fixed point. It is similar, and therefore isomorphic to the group of all translations $Z^* = z^* + b$. The pencils of invariant and interchanged circles associated with this group are two orthogonal parabolic pencils passing through γ. All parabolic Moebius transformations which are products of inversions within one of these pencils form a subgroup $\mathscr{H}_p^{(1)} = \mathscr{H}_p^{(1)}(\gamma)$. Similarly there is a subgroup $\mathscr{H}_p^{(2)} = \mathscr{H}_p^{(2)}(\gamma)$ of products of inversions with respect to the other pencil. Then $\mathscr{H}_p$ is the direct product of $\mathscr{H}_p^{(1)}$ and $\mathscr{H}_p^{(2)}$.

Remark. Using the normal forms of the pencils, the following facts are easily established. The group $\mathscr{H}_e$ is isomorphic to the group of all plane rotations about a fixed point, or to the addition group of the real numbers (mod 2π). The group $\mathscr{H}_p^{(1)}$ is isomorphic to the group of all translations in a given direction, hence to the addition group of all real numbers. The group $\mathscr{H}_h^+$ is isomorphic to the multiplication group of all positive real numbers, and the group $\mathscr{H}_h$ is isomorphic to the multiplication group of all non-zero real numbers (cf. § 10, ex. 12).

Examples

1. Let $\mathfrak{K} = \begin{pmatrix} a & b \\ c & d \end{pmatrix}$ be a circle matrix. As in the proof of Theorem A put

$$c = |c|e^{i\phi} \quad \text{if} \quad c \neq 0.$$

We shall then call

$$\mathfrak{K}_1 = ie^{-i\phi}\mathfrak{K} = \begin{pmatrix} a_1 & ib_1 \\ ic_1 & \bar{a}_1 \end{pmatrix} \tag{9.5}$$

a 'normalized circle matrix.' Here b_1 and c_1 are real and a_1 complex. The condition $c \neq 0$ is not essential for the normalization; for every circle matrix $\mathfrak{K}$ a suitable factor of modulus one can be found such that $\mathfrak{K}_1 = \lambda\mathfrak{K}$ has the form (9.5). Normalization does not change the multiplier: $\mu = |\mathfrak{K}_1|$.

Making use of induction with respect to n show that the powers $\mathfrak{K}_1^n$ of a normalized circle matrix $\mathfrak{K}_1$ are also normalized circle matrices.

2. Show that the circles $\mathfrak{C}_1$, $\mathfrak{C}_2$ associated with the (normalized) circle matrices $\mathfrak{K}_1$, $\mathfrak{K}_2$ are orthogonal if and only if the inversions $z_1 = \mathfrak{K}_1(\bar{z})$, $z_2 = \mathfrak{K}_2(\bar{z})$ are commutative:

$$\mathfrak{K}_1\overline{\mathfrak{K}}_2 + \mathfrak{K}_2\overline{\mathfrak{K}}_1 = 0. \qquad \text{(cf. § 1, ex. 8.)}$$

3. The class of all anti-homographies $Z = \mathfrak{H}(\bar{z})$ for which the square $\mathfrak{H}\overline{\mathfrak{H}}$ represents an involution, has its normal form the transformation

$$Z^* = \frac{i}{\overline{z^*}} \qquad (\beta = \pi).$$

Every element of this class is defined by its conjugate pair γ_1, γ_2. If these are finite points, the corresponding anti-homography $Z = \mathfrak{H}(\bar{z})$ has as matrix

$$\mathfrak{H} = \begin{pmatrix} \gamma_1 - i\gamma_2 & -\gamma_1\bar{\gamma}_1 + i\gamma_2\bar{\gamma}_2 \\ 1-i & -\bar{\gamma}_1 + i\bar{\gamma}_2 \end{pmatrix}.$$

If one point of the pair, γ_2, is infinite, one finds

$$\mathfrak{H} = \begin{pmatrix} \gamma_1 & i - \gamma_1\bar{\gamma}_1 \\ 1 & -\bar{\gamma}_1 \end{pmatrix}.$$

4. The system of all Moebius transformations and all anti-homographies is a group in which the group of all Moebius transformations is subgroup of index two.

5. The system of all Moebius transformations which either preserve, or interchange, two given distinct points γ_1, γ_2, is a non-commutative group

$$\hat{\mathscr{H}} = \mathscr{H} + \mathfrak{J}\mathscr{H}$$

where $\mathfrak{J}$ is an involution having γ_1, γ_2 as a conjugate pair; for example,

$$\mathfrak{J}(z) = -z + \gamma_1 + \gamma_2.$$

All elements of the coset $\mathfrak{J}\mathscr{H}$ are involutions interchanging γ_1 and γ_2.

6. The group of *all* Moebius transformations $\mathfrak{H}$ having γ_1 as fixed point is similar, and therefore isomorphic, to the group of all integral Moebius transformations. For finite γ_1:

$$\mathfrak{H} = \begin{pmatrix} b\gamma_1 + 1 & \gamma_1(a - b\gamma_1 - 1) \\ b & a - b\gamma_1 \end{pmatrix} \qquad (a \neq 0).$$

The second fixed point of this transformation will be

$$\gamma_2 = \gamma_1 + \frac{1-a}{b}.$$

7. *Similarity classification of the Moebius transformations with circle matrices.* Let

$$\mathfrak{K} = \begin{pmatrix} a & b_2 i \\ c_2 i & \bar{a} \end{pmatrix} \qquad (a = a_1 + ia_2;\ a_1, a_2, b_2, c_2 \text{ real})$$

be a normalized circle matrix and $\mu = |\mathfrak{K}| = a_1^2 + a_2^2 + b_2c_2$ its multiplier. The similarity class of $\mathfrak{K}$ is defined by the value

$$\sigma = \sigma(\mathfrak{K}) = \frac{\tau^2}{\mu} - 4 \qquad (\tau = \operatorname{tr} \mathfrak{K} = 2a_1) \tag{9.6}$$

$$= -4\,\frac{a_2^2 + b_2c_2}{a_1^2 + a_2^2 + b_2c_2}$$

which is real; hence (§ 8, **c**) $\mathfrak{R}$ cannot be loxodromic. The fixed points of $\mathfrak{R}$ satisfy the quadratic equation with real coefficients

$$c_2 z^2 - 2a_2 z - b_2 = 0$$

whose discriminant is $4(a_2^2 + b_2 c_2)$.

There are the following possibilities

(i) $\mathfrak{R}$ improper hyperbolic: $\sigma \leqslant -4$. If $\sigma < -4$ then necessarily $\mu < 0$; hence $a_2^2 + b_2 c_2 < 0$, and thus γ_1, γ_2, the fixed points of $Z = \mathfrak{R}(z)$, are non-real and conjugate complex. If $\mathfrak{R}$ represents an involution ($a_1 = 0$) and $\mu < 0$, we may write $\sigma = -4 - 0$.

(ii) $\mathfrak{R}$ elliptic: $-4 \leqslant \sigma < 0$; then $0 < a_2^2 + b_2 c_2 \leqslant \mu$ and γ_1, γ_2 are real. An elliptic involution ($a_1 = 0$, $\mu > 0$) may be characterized by $\sigma = -4+0$.

(iii) $\mathfrak{R}$ parabolic: $\sigma = 0$, that is, $a_2^2 + b_2 c_2 = 0$, $\mu = a_1^2 > 0$ and the fixed point γ is real.

(iv) $\mathfrak{R}$ proper hyperbolic: $\sigma > 0$, thus $a_2^2 + b_2 c_2 < 0$, $\mu > 0$ and the fixed points γ_1, γ_2 are non-real and conjugate complex.

The normal form $\mathfrak{T}\mathfrak{R}\mathfrak{T}^{-1}$ can be produced by a real transformation $\mathfrak{T}$. It is readily seen that the matrix of an involution is $\mathfrak{R} = i \begin{pmatrix} a_2 & b_2 \\ c_2 & -a_2 \end{pmatrix}$; the involution is therefore real (cf. § 8, **d**).

The character of the Moebius transformation $Z = \mathfrak{R}(z)$ may be derived from the position of the circle $\mathfrak{C}$, given by the equation (9.33), viz. $z = \mathfrak{R}(\bar{z})$, with respect to the real axis defined by $z = \bar{z}$. If the intersections of $\mathfrak{C}$ with the real axis are real, they are the fixed points of $Z = \mathfrak{R}(z)$. The circle $\mathfrak{C} = \begin{pmatrix} c_2 & -\bar{a}i \\ ai & -b_2 \end{pmatrix}$ ($|\mathfrak{C}| = -\mu$) has $z = -ai/c_2$ as centre and $\rho = \sqrt{(\mu)}/|c_2|$ as radius. The distance of its centre from the real axis is $|a_1/c_2|$. Hence in the cases listed above

(i) $\mathfrak{C}$ is an imaginary circle.

(ii) $\mathfrak{C}$ intersects the real axis in two real points (and has its centre on the real axis if $\mathfrak{R}$ represents an involution).

(iii) $\mathfrak{C}$ touches the real axis.

(iv) $\mathfrak{C}$ has no common point with the real axis: $a_1^2/c_2^2 > \mu/c_2^2$.

If $c_2 = 0$ the circle $\mathfrak{C}$ is the line $-a_2 x + a_1 y = b_2/2$.

A special set of Moebius transformations with circle matrices $\mathfrak{R} = \mathfrak{H}_\rho$ has been studied in § 7, ex. 4. The associated circles $\mathfrak{C}$ (not the circles $\tilde{\mathfrak{C}}_\rho$) are concentric with the centre $z = i$. Cf. also § 8, ex. 7.

8. A Moebius transformation $Z = \mathfrak{R}(z)$ with the circle matrix $\mathfrak{R} = \begin{pmatrix} -C & -D \\ A & B \end{pmatrix}$ ($A \neq 0$) has as 'inverse pole' $Z_\infty = \mathfrak{R}(\infty)$ the centre of the fundamental circle $\mathfrak{C}$ of the inversion $z^* = \mathfrak{R}(\bar{z})$. Its own pole is $z_\infty = \bar{Z}_\infty$. Its fixed points are the real or non-real intersections γ_1, γ_2 of $\mathfrak{C}$ with the real axis, that is, the roots of the equation

$$Az^2 + (B+C)z + D = 0. \tag{9.7}$$

Thus if γ_1, γ_2 are not real, $\mathrm{Re}(\gamma_1) = \mathrm{Re}(Z_\infty)$. The different similarity types are distinguished by their characteristic parallelograms. All the possibilities are shown in Figure 23.

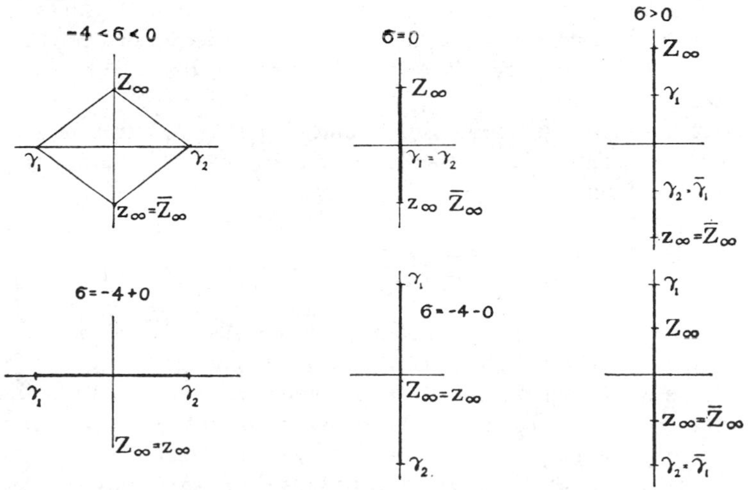

FIG. 23

§ 10. Iteration of a Moebius transformation

a. *General remarks on iteration.* 'Iteration' means 'repeated application of a mathematical operation.' It plays an important role in almost all parts of mathematics. For example, the power c^n of a number c is obtained by n-fold iteration of multiplication by c.

Let $f(z)$ be a function defined within a certain domain $\mathscr{D}$ of the z-plane that may be the whole completed plane. For every point z_0 of $\mathscr{D}$ let $f(z_0) = z_1$ also be a point of $\mathscr{D}$. Thus the value $f(z_1) = z_2$ can again be formed and is a point in $\mathscr{D}$. This process may be repeated any number of times. By iteration of the functional operation $f(z)$ applied to the initial value z_0 we obtain in this way the 'recurrent sequence'

$$z_0,\ z_1 = f(z_0), \ldots,\ z_n = f(z_{n-1}) \qquad (n = 1, 2, \ldots). \tag{10.1}$$

If $f(z)$ is a continuous function in $\mathscr{D}$ and if the sequence (10.1) has a *limit* γ as $n \to \infty$, then γ is necessarily a *fixed point* of the function $f(z)$, that is, $f(\gamma) = \gamma$. In fact

$$f(\lim z_n) = \lim f(z_n) = \lim z_{n+1} = \lim z_n.$$

A fixed point γ is said to be *attractive* if for all z_0 in a certain neighbourhood of γ the recurrent sequence (10.1) is convergent and has γ as its limit.

The function $f(z) = kz$ $(k \neq 0)$ has 0 and ∞ as fixed points. If $|k| < 1$ the fixed point 0 is attractive since $z_n = k^n z_0 \to 0$ for every complex z_0. If $|k| > 1$, then the fixed point ∞ will be attractive since $k^n z_0 \to \infty$ for every $z_0 \neq 0$. In each case the points z_n of the sequence, while approaching the attractive fixed point, run away from the other fixed point which therefore is said to be *repulsive*.

If $|k| = 1$ the points $k^n z_0$ with increasing n move on the circle with the radius $|z_0|$ about the fixed point 0. Thus they approach neither the one nor the other fixed point. In this case the fixed points of the function kz are said to be *indifferent*.

The property of a fixed point γ of a function $f(z)$ to be either attractive or repulsive or indifferent, is invariant if a new variable z^* is introduced by a Moebius transformation $z^* = \mathfrak{T}(z)$ (cf. § 8, **a**, **b**). Indeed if $z_n \to \gamma$, then $\mathfrak{T}(z_n) \to \mathfrak{T}(\gamma)$, the fixed point of the function $f^*(z^*) = \mathfrak{T}\{f[\mathfrak{T}^{-1}(z^*)]\}$; and

$$\mathfrak{T}(z_n) = \mathfrak{T}[f(z_{n-1})] = \mathfrak{T}(f\{\mathfrak{T}^{-1}[\mathfrak{T}(z_{n-1})]\})$$

is the recurrent sequence of the function $f^*(z^*)$ for the initial value $z_0^* = \mathfrak{T}(z_0)$. If the z_n move away from γ, then the $\mathfrak{T}(z_n)$ cannot approach $\mathfrak{T}(\gamma)$.

It may be noted that these statements remain true if a new variable z^* is introduced by any (not necessarily Moebius) transformation $z^* = \varphi(z)$ which is a topological mapping of $\mathscr{D}$ onto itself (cf. § 3, **a**).

If we consider the recurrent sequence (10.1) of the function $f(z)$ with variable initial term z in $\mathscr{D}$, then we are really concerned with the sequence of functions

$$f^0(z) = z, \quad f^1(z) = f(z), \quad f^2(z) = f[f(z)], \ldots, \quad f^n(z) = f[f^{n-1}(z)], \ldots, \tag{10.2}$$

which we call the *iterates of the function* $f(z)$. The notation used presently suggests considering the successive iterates of $f(z)$ as 'powers' of the operation f applied to the variable z.

By (10.2) the iterates (powers) of f are defined for all non-negative integral exponents (indices) n. If for all z in $\mathscr{D}$ the function $f(z)$ is invertible and its (one-valued) inverse is denoted by $f^{-1}(z)$, we may extend the definition of $f^n(z)$ to negative exponents:

$$f^{-m}(z) = (f^{-1})^m(z) \qquad (m > 0). \tag{10.21}$$

It is readily seen then that for all integral m, n the usual index laws are valid:

$$f^m[f^n(z)] = f^{m+n}(z), \qquad (f^m)^n(z) = f^{mn}(z). \tag{10.22}$$

If γ is a fixed point of $f(z)$ then it is also a fixed point for all iterates $f^n(z)$ $(n > 0)$. It is an attractive, repulsive, or indifferent fixed point of $f^n(z)$ according as it is an attractive, repulsive, or indifferent fixed point of $f(z)$. If $f^{-1}(z)$ exists, it will still have γ as a fixed point, but possibly of different character, as for instance in the case $f(z) = kz$.

b. *Iteration of a Moebius transformation.* If $Z = f(z) = \mathfrak{H}(z)$ is a Moebius transformation, the domain $\mathscr{D}$ will be the completed plane. All iterates of $\mathfrak{H}$ (with positive as well as negative exponents) are Moebius transformations; their matrices are the powers $\mathfrak{H}^n$ of the matrix $\mathfrak{H}$, and a recurrent sequence will be defined for every initial point z_0, viz.

$$z_n = \mathfrak{H}^n(z_0). \tag{10.3}$$

The convergence behaviour of this sequence will be the same as that of the sequence

$$z_n^* = \mathfrak{T}(z_n) = (\mathfrak{T}\mathfrak{H}\mathfrak{T}^{-1})^n\mathfrak{T}(z_0). \tag{10.31}$$

Therefore, if we choose $\mathfrak{T}$ so that $\mathfrak{T}\mathfrak{H}\mathfrak{T}^{-1}$ is the normal form of $\mathfrak{H}$ (cf. § 8, **b**, **c**), the sequence z_n^* will be given by

$$z_n^* = k^n z_0^* \qquad \text{if } \mathfrak{H} \text{ is non-parabolic,}$$

$$z_n^* = z_0^* + n \qquad \text{if } \mathfrak{H} \text{ is parabolic.}$$

Hence

Theorem A. *The fixed point of a parabolic Moebius transformation is always attractive. Of the two different fixed points of a non-parabolic Moebius transformation $\mathfrak{H}$ the one is attractive, the other is repulsive, except when $\mathfrak{H}$ is elliptic. The fixed points of an elliptic transformation are both indifferent.*[8]

Corollary. *The sequence* (10.3) *is always convergent, and has as its limit the attractive fixed point of $\mathfrak{H}$, if $\mathfrak{H}$ is not elliptic and if the initial point z_0 of the sequence is not the repulsive fixed point of $\mathfrak{H}$.*

Here we consider a sequence as convergent even when its limit is ∞. This convention is not generally accepted in the theory of functions of a complex variable where the point at infinity often plays an exceptional role. It is, however, useful in the geometry of the completed plane of the complex numbers where ∞ plays the role of an ordinary point; its naturalization has been justified by the stereographic projection of the plane onto the sphere.[9]

If $\mathfrak{H}$ has an attractive and a repulsive fixed point, γ_1 and γ_2 respectively, then γ_2 is the attractive, γ_1 the repulsive fixed point of the transformation $\mathfrak{H}^{-1}$.

c. *Periodic sequences of Moebius transformations.* Let $\mathfrak{H}_1, \mathfrak{H}_2, \ldots$, be a sequence of Moebius transformations. For any given initial point z_0 we consider the sequence

$$z_0, \; z_1 = \mathfrak{H}_1(z_0), \qquad z_2 = \mathfrak{H}_1[\mathfrak{H}_2(z_0)] \ldots . \tag{10.4}$$

[8] In this theorem every involution ($k = -1$) is to be considered as elliptic.

[9] If homogeneous coordinates are used (cf. § 6, a, Remark 1) the point at infinity is, like all the other points of the completed plane, given by a pair of complex numbers, not both equal to zero, viz. the pair (1, 0) (cf. § 5, **b**, Remark).

The question of the convergence of such a sequence appears to be a difficult one.[10] The preceding Theorem A, however, enables us to decide it in the important case where the given sequence of transformations $\mathfrak{H}_n$ is *periodic* which means that there is a least integer $p \geqslant 1$, *the period* of the sequence, such that the sequence of the corresponding matrices may be written in the form

$$\mathfrak{H}_1, \mathfrak{H}_2, \ldots, \mathfrak{H}_p, \mathfrak{H}_1, \mathfrak{H}_2, \ldots, \mathfrak{H}_p, \ldots,$$

so that

$$\mathfrak{H}_{mp+r} = \mathfrak{H}_r \qquad (0 < r \leqslant p;\, m = 0, 1, 2, \ldots). \tag{10.41}$$

We introduce the *period matrix* of the sequence

$$\mathfrak{H} = \mathfrak{H}_1 \mathfrak{H}_2 \ldots \mathfrak{H}_p. \tag{10.42}$$

If $n = mp+r$ $(0 < r \leqslant p)$, then

$$z_n = \mathfrak{H}^m \mathfrak{H}_1 \ldots \mathfrak{H}_r(z_0) = \mathfrak{H}^m(z_r).$$

Suppose that none of the p values $z_0, z_1, \ldots, z_{p-1}$, viz.

$$z_0, \quad z_r = \mathfrak{H}_1 \mathfrak{H}_2 \ldots \mathfrak{H}_r(z_0)$$

coincides with the repulsive fixed point of $\mathfrak{H}$. If $\mathfrak{H}$ is not elliptic, then the sequences

$$z_{mp} = \mathfrak{H}^m(z_0), \qquad z_{mp+1} = \mathfrak{H}^m(z_1), \ldots, z_{mp+p-1} = \mathfrak{H}^m(z_{p-1})$$

all tend to the same limit γ_1 (the attractive fixed point of $\mathfrak{H}$) as $m \to \infty$. So, therefore, does the sequence (10.4) whose elements are distributed on these p sequences. None of these sequences is convergent if $\mathfrak{H}$ is elliptic, except when one of the z_r is a fixed point of $\mathfrak{H}$. Hence the following

Theorem B. *A sequence* (10.4) *associated with a periodic sequence of Moebius transformations* $\mathfrak{H}_1, \mathfrak{H}_2, \ldots,$ *with the period* $p \geqslant 1$ *is convergent if and only if*

Either: The first p *values* $z_0, z_1, \ldots, z_{p-1}$ *are all equal to the same fixed point of the Moebius transformation with the period matrix* $\mathfrak{H}$.

Or: $\mathfrak{H}$ *has an attractive fixed point* γ_1 *(that is,* $\mathfrak{H}$ *is non-elliptic and* $\mathfrak{H} \neq \mathfrak{E}$*) and none of the points* $z_0, z_1, \ldots, z_{p-1}$ *coincides with the repulsive fixed point of* $\mathfrak{H}$ *if there is one. Then*

$$\lim_{n \to \infty} z_n = \gamma_1.$$

An important application of Theorem B in the theory of continued fractions will be discussed in ex. 2.

d. *Moebius transformations with periodic iteration.* The Moebius transformation $\mathfrak{H}$ is said to have a *periodic iteration of length p* if there is a complex number z_0 such that

$$\mathfrak{H}(z_0) = z_1, \quad \mathfrak{H}(z_1) = z_2, \ldots, \quad \mathfrak{H}(z_{p-1}) = z_0 \tag{10.5}$$

[10] Cf. G. Piranian and W. J. Thron [1], and P. Erdös and G. Piranian [1].

and the numbers $z_0, z_1, \ldots, z_{p-1}$ are all different. They form an 'iteration period' of $\mathfrak{H}$. If $\mathfrak{T}$ is another Moebius transformation, then $\mathfrak{T}(z_0), \mathfrak{T}(z_1), \ldots, \mathfrak{T}(z_{p-1})$ represents an iteration period of the transformation $\mathfrak{T}\mathfrak{H}\mathfrak{T}^{-1}$.

If $p = 1$, then $z_1 = z_0$ is one of the fixed points of $Z = \mathfrak{H}(z)$. Conversely, a fixed point produces a periodic iteration of length one, for any $\mathfrak{H}$ whatsoever.

Let $p > 1$. If $\mathfrak{H}$ is parabolic it is similar to the translation $\begin{pmatrix}1 & 1\\ 0 & 1\end{pmatrix}$ whose nth iteration is $\begin{pmatrix}1 & n\\ 0 & 1\end{pmatrix}$. Thus $\mathfrak{H}$ admits no iterations of period $p > 1$.

If $\mathfrak{H}$ is non-parabolic and different from the identity it is similar to $\begin{pmatrix}k & 0\\ 0 & 1\end{pmatrix}$ $(k \neq 1)$, and $\mathfrak{H}^p$ is similar to $\begin{pmatrix}k^p & 0\\ 0 & 1\end{pmatrix}$. Therefore $\mathfrak{H}^p(z_0) = z_0$ for $z_0 \neq \gamma$, implies

$$k^p z_0^* = z_0^* \quad (z_0^* \neq 0), \qquad k^p = 1.$$

Thus $\mathfrak{H}$ is elliptic and its characteristic constant k is a primitive pth root of unity:

$$k = e^{(2\pi i\nu)/p} \qquad (\nu,\, p \text{ relatively prime}). \tag{10.51}$$

Hence

Theorem C. *If an iteration of a Moebius transformation $\mathfrak{H}$ is periodic of length $p > 1$, then $\mathfrak{H}$ is elliptic and of order p (that is, p is the least positive integer such that $\mathfrak{H}^p = \mathfrak{E}$).*

We notice that for $p = 2$ the transformation $\mathfrak{H}$ is an involution (cf. § 6, **f**, Theorem F and § 6, ex. 10). Another proof of Theorem C will be indicated in ex. 4.

According to Theorem C we may direct our attention to the Moebius transformations $\mathfrak{H}$ of finite order $p \geqslant 2$. By (8.45) and (10.51)

$$\sigma = k + \frac{1}{k} - 2 = -4\left(\sin\frac{\nu\pi}{p}\right)^2 = 2\left(\cos\frac{2\nu\pi}{p} - 1\right) \tag{10.52}$$

whence for small p we obtain the following table:

(10.53)

p	2	3	4	6
σ	-4	-3	-2	-1

Since for an elliptic $\mathfrak{H}$ we always have $-4 \leqslant \sigma < 0$, we conclude that there are no values of p different from 2, 3, 4, 6 such that $\sigma = \sigma(\mathfrak{H})$ has integral value. Indeed, Moebius transformations of different orders cannot be similar and therefore belong to different values of the invariant σ.

The points $z_0, z_1, \ldots, z_{p-1}$ of a periodic iteration are always situated on an invariant circle of the transformation $\mathfrak{H}$. By a Moebius transformation $z^* = \mathfrak{T}(z)$

which carries three different points of this circle into three points of the real axis in the z^*-plane, we obtain the transformation $\mathfrak{T}\mathfrak{H}\mathfrak{T}^{-1}$ for which the real axis in the z^*-plane is the invariant circle on which lie the points $\mathfrak{T}(z_0)$, $\mathfrak{T}(z_1), \ldots,$ $\mathfrak{T}(z_{p-1})$ of a real periodic iteration of the transformation $\mathfrak{T}\mathfrak{H}\mathfrak{T}^{-1}$. This transformation is evidently real in the sense of § 8, **d.** Thus we may assume that the matrix $\mathfrak{H}$ is real, that is, its elements are real numbers.

Moreover we shall now suppose that these elements are *rational numbers.* Then $\mathrm{tr}\,\mathfrak{H}$, $|\mathfrak{H}|$, $\sigma(\mathfrak{H})$ and $k+k^{-1}$ are rational. Thus either k is rational or k is the root of a quadratic equation with rational coefficients. On the other hand $k^p = 1$ and $k^n \neq 1$ for $0 < n < p$. From algebra it is known that k then is a root of an irreducible polynomial with rational coefficients, a divisor of the polynomial $x^p - 1$. The degree of this irreducible polynomial (the so-called pth cyclotomic polynomial) is equal to $\phi(p)$, that is, the number of positive integers less than p and prime to p. Because the degree of an irreducible equation of which a given algebraic number is a root, is uniquely defined by this number, we conclude that $\phi(p)$ must be equal to 1 or to 2. From elementary number theory it is known then that p cannot have any other value except 2, 3, 4, 6. Thus

Theorem D. *The only possible finite orders of a Moebius transformation with rational coefficients are* $p = 1, 2, 3, 4, 6$.

e. *Continuous iteration.* We shall now extend the definition of the functional powers (iterates, cf. (10.2)) from integral exponents n to arbitrary real exponents s. Let $f(s, z)$ be a function which for all values of the real parameter s is defined for all z in $\mathcal{D}$ and has its values in $\mathcal{D}$. We shall call it a continuous (or analytic) iteration of the function $f(z)$, and denote it by

$$f(s, z) = f^s(z) \tag{10.6}$$

if it satisfies the following conditions:

$$f(0, z) = z, \qquad f(1, z) = f(z) \tag{10.61}$$

$$f[s_2, f(s_1, z)] = f(s_1+s_2, z) \tag{10.62}$$

for all z in $\mathcal{D}$ and for all real s_1, s_2. Such a family of functions $f^s(z)$ represents a one-parameter group of transformations. Thus the problem of continuous iteration of a given function (or transformation) $Z = f(z)$ amounts to the following question: is it possible to determine a one-parameter group of transformations $Z = f(s, z)$ which contains the given transformation as an element?

In order to proceed to a partial solution of this problem let us assume that the function $f(z)$ for z in $\mathcal{D}$ is similar to the function kz^* where k is a complex constant. This means that for all z in $\mathscr{D}$ a function $z^* = \phi(z)$ with a unique inverse $z = \phi^{-1}(z^*)$ exists (cf. ex. 7) such that

$$f(z) = \phi^{-1}[k\phi(z)]. \tag{10.63}$$

This equation, often written in the form

$$\phi[f(z)] = k\phi(z) \tag{10.64}$$

is called Schröder's functional equation. Assuming that it has a solution $\phi(z)$ we can easily verify that

$$f(s, z) = \phi^{-1}[k^s\phi(z)] \tag{10.65}$$

represents a family of functions that satisfies the conditions (10.61) and (10.62), that is, a continuous iteration of $f(z)$. We notice that other continuous iterations of $f(z)$ are given by

$$f_m(s, z) = \phi^{-1}[e^{2\pi ims}k^s\phi(z)] \qquad (m = 0, \pm 1, \pm 2, \ldots). \tag{10.66}$$

The continuous iteration problem has a solution also if the given function $f(z)$ is similar to a translation $Z^* = z^*+1$. By this supposition there is a function $\psi(z) = z^*$, defined in $\mathscr{D}$, with a unique inverse, so that

$$f(z) = \psi^{-1}[\psi(z)+1]. \tag{10.67}$$

This relation, also written in the form $\psi(f(z)) = \psi(z)+1$, is known as Abel's functional equation. In this case

$$f(s, z) = \psi^{-1}[\psi(z)+s] \tag{10.68}$$

represents a continuous iteration of the function $f(z)$.[11]

f. *Continuous iteration of a Moebius transformation.* In the case of a Moebius transformation $Z = \mathfrak{H}(z)$ the continuous iteration problem can always be solved by means of a family (group) of Moebius transformations.

1. Let $\mathfrak{H}$ be non-parabolic, k its characteristic constant, and γ_1, γ_2 its fixed points, both finite. By solving the equation (8.41) with respect to Z we find the following representation of the transformation $\mathfrak{H}$ which sets the constant k in evidence:

$$Z = \mathfrak{H}(z) = \frac{(k\gamma_2-\gamma_1)z+(1-k)\gamma_1\gamma_2}{(k-1)z+\gamma_2-k\gamma_1}. \tag{10.7}$$

[11] It may be pointed out that, in a formal way, the two functions kz and $z+1$ which are characteristic for the Schröder and the Abel case respectively, are 'similar' to each other. In fact for $f(z) = z+1$ Schröder's equation (10.64) is solved by the function $\phi(z) = k^z$. However, this function does not possess a unique (one-valued) inverse. The formal 'similarity' is therefore not pertinent; it would, moreover, dispose of the essential constant k in Schröder's equation. (This remark should obviate the statement frequently found in the literature to the effect that Schröder's and Abel's functional equations are 'equivalent.' Cf. for example, E. Picard [1], p. 155.)

We introduce now the two matrices, of rank one:

$$\mathfrak{L}_1 = \begin{pmatrix} \gamma_2 & -\gamma_1\gamma_2 \\ 1 & -\gamma_1 \end{pmatrix}, \qquad \mathfrak{L}_2 = \begin{pmatrix} \gamma_1 & -\gamma_2\gamma_1 \\ 1 & -\gamma_2 \end{pmatrix}$$

of which one arises from the other by interchanging γ_1 and γ_2. Then (10.7) can be written in the form

$$\mathfrak{H} = q(k\mathfrak{L}_1 - \mathfrak{L}_2) \tag{10.71}$$

with an arbitrary complex factor $q \neq 0$.

According to (10.65) the continuous iteration of the transformation $\mathfrak{H}$ is obtained by formally replacing k by k^s in (10.7). A matrix of the Moebius transformation (as well as the transformation itself) which in this family belongs to the value s of the parameter, may be denoted by $\mathfrak{H}^s$. Hence

$$\mathfrak{H}^s = q(s)(k^s\mathfrak{L}_1 - \mathfrak{L}_2) \tag{10.72}$$

where $q(s)$ is an arbitrary function of the real parameter s; thus we may put $q(s) \equiv 1$. It should be noted that (10.72) is a relation between matrices.

For integral $s = n$ the formula (10.72) defines uniquely the nth iteration of the transformation $\mathfrak{H}$ (cf. ex. 8). In general, however, the transformation $\mathfrak{H}^s$ is not unique. Since $k^s = e^{s \log k}$ it depends on the value chosen for $\log k$. If $k = \rho e^{\alpha i}$, then

$$\log k = \log \rho + i(\alpha + 2m\pi) \qquad (\rho = |k|)$$

where m may be any integer. In most cases it will be natural to choose the so-called principal value of $\log k$, that is, $m = 0$ if α is restricted to the interval $-\pi < \alpha < \pi$. If $k > 0$, the principal value is the elementary logarithm of a positive number.

If $|k| < 1$, then the matrix equation (10.72) enables us to obtain the limit

$$\lim_{n\to\infty} \mathfrak{H}^n = -\mathfrak{L}_2, \tag{10.721}$$

that is, a matrix of rank one which does not determine a Moebius transformation in the sense of the original definition. Extending, however, the usual symbolism by which a matrix determines a Moebius transformation, we find

$$\mathfrak{L}_2(z) = \frac{\gamma_1 z - \gamma_2\gamma_1}{z - \gamma_2} = \gamma_1 \qquad \text{if } z \neq \gamma_2 \tag{10.722}$$

whereas for $z = \gamma_2$ this 'degenerated Moebius transformation' is not defined. Thus we call γ_2 its 'indefinite point.' We notice that the last formula yields another proof of the first part of Theorem A.

2. Let $\mathfrak{H}$ be parabolic and γ its fixed point. By the transformation

$$z^* = \mathfrak{T}(z) = \frac{h}{z-\gamma}$$

with a suitable $h \neq 0$ the transformation $\mathfrak{H}$ will be taken into $\mathfrak{T}\mathfrak{H}\mathfrak{T}^{-1}(z^*) = z^*+1$, and by (10.67)

$$\mathfrak{H}(z) = \mathfrak{T}^{-1}[\mathfrak{T}(z)+1],$$

whence by (10.68)

$$\mathfrak{H}^s(z) = \frac{\gamma\left(\dfrac{h}{z-\gamma}+s\right)+h}{\dfrac{h}{z-\gamma}+s} = \frac{(h+\gamma s)z-\gamma^2 s}{sz+h-\gamma s}.$$

In terms of matrices,

$$\mathfrak{H}^e = q(s)(h\mathfrak{E}+s\mathfrak{L}), \tag{10.73}$$

where $\mathfrak{L} = \begin{pmatrix} \gamma & -\gamma^2 \\ 1 & -\gamma \end{pmatrix}$ again represents a degenerated Moebius transformation which, however, unlike $\mathfrak{L}_2$, has no fixed point since $\mathfrak{L}(z) = \gamma$ for all $z \neq \gamma$ and γ is its indefinite point. The function $q(s)$ is arbitrary (cf. ex. 9).

With every non-loxodromic Moebius transformation $\mathfrak{H}$ we had associated a pencil of invariant circles and a pencil of interchanged circles which in the non-parabolic cases are uniquely defined by the two fixed points γ_1, γ_2 of $\mathfrak{H}$. If $\mathfrak{H}$ is proper hyperbolic, all $\mathfrak{H}^s$ are proper hyperbolic and have the same fixed points and therefore the same pair of orthogonal pencils. If $\mathfrak{H}$ is elliptic, all $\mathfrak{H}^s$ are also elliptic (for any determination of log k). Every involution is taken to be elliptic. If $\mathfrak{H}$ is improper hyperbolic, then $\alpha = \pi$ and for non-integral values of s the transformation $\mathfrak{H}^s$ is loxodromic. Finally if $\mathfrak{H}$ is parabolic, all $\mathfrak{H}^s$ are parabolic. Indeed by (10.73)

$$\operatorname{tr} \mathfrak{H}^s = 2qh, \qquad |\mathfrak{H}^s| = q^2h^2;$$

thus

$$\sigma(\mathfrak{H}^s) = 0.$$

Now let $\mathfrak{M}_1 = \begin{pmatrix} a & b \\ c & d \end{pmatrix}$, $(c \neq 0)$ be a matrix of rank one such that γ_1 is the indefinite, γ_2 the fixed point of the corresponding degenerated Moebius transformation:

$$\mathfrak{M}_1(z) = \frac{az+b}{cz+d} = \gamma_2 \qquad \text{for all } z \neq \gamma_1.$$

It will be shown that $\mathfrak{M}_1$ is a numerical multiple of $\mathfrak{L}_1$. Indeed by hypothesis

$$az+b = \gamma_2(cz+d), \quad \text{or} \quad a = \gamma_2 c, \qquad b = \gamma_2 d$$

and also

$$c\gamma_1+d = 0, \quad \text{that is,} \quad d = -\gamma_1 c.$$

Hence

$$\mathfrak{M}_1 = c\mathfrak{L}_1.$$

Similarly the matrix $\mathfrak{L}$ with the indefinite point γ and no fixed point is defined apart from a factor.

Now it is readily seen that if $\mathfrak{L}_1$ is a matrix of rank one with indefinite point γ_1 and fixed point γ_2 then the matrix $\mathfrak{T}\mathfrak{L}\mathfrak{T}^{-1}$ has rank one, the indefinite point $\mathfrak{T}(\gamma_1)$ and the fixed point $\mathfrak{T}(\gamma_2)$. Hence

$$\mathfrak{T}\mathfrak{L}_1\mathfrak{T}^{-1} = q\begin{pmatrix} \mathfrak{T}(\gamma_2) & -\mathfrak{T}(\gamma_1)\mathfrak{T}(\gamma_2) \\ -1 & -\mathfrak{T}(\gamma_1) \end{pmatrix}, \tag{10.74}$$

with corresponding relations for $\mathfrak{L}_2$ and $\mathfrak{L}$, for every (non-degenerated) Moebius transformation $\mathfrak{T} = \begin{pmatrix} t & u \\ v & w \end{pmatrix}$ for which γ_1 and γ_2 are not poles. The factor q in the matrix equation (10.74) is found to be

$$q = \frac{1}{|\mathfrak{T}|}(v\gamma_1 + w)(v\gamma_2 + w); \tag{10.75}$$

it is symmetric in γ_1, γ_2. It will therefore remain the same when $\mathfrak{L}_1$ is replaced by $\mathfrak{L}_2$.

Thus with regard to (10.72) we conclude that

$$\mathfrak{T}\mathfrak{H}^s\mathfrak{T}^{-1} = (\mathfrak{T}\mathfrak{H}\mathfrak{T}^{-1})^s \qquad \text{for all real } s, \tag{10.76}$$

which is obvious if s is an integer.

If $\mathfrak{H}$ is parabolic the relation (10.76) is still valid, which can conveniently be seen by normalizing the matrix $\mathfrak{T}$ in such a way that $(v\gamma + w)^2 = 1$. Then (10.76) follows immediately from (10.73).

With respect to the normal forms and to (10.76) the following theorem will become evident. It provides a kinematical interpretation of the continuous iteration of a Moebius transformation.

Theorem E. *Let $\mathfrak{H}$ be a proper hyperbolic or parabolic or elliptic Moebius transformation and z_0 a constant point which is not a fixed point of $\mathfrak{H}$. Then the point*

$$z = \mathfrak{H}^s(z_0)$$

moves on the invariant circle $\mathfrak{C}$ of $\mathfrak{H}$ through z_0.

Corollary 1. *If $\mathfrak{H}$ is elliptic (including the case of an involution) then the point z describes the complete circle $\mathfrak{C}$ infinitely often if s runs through all real numbers. If $\mathfrak{H}$ is proper hyperbolic (and thus the pencil of invariant circles elliptic) then the point z remains always on the arc, between the fixed points of $\mathfrak{H}$, on which the initial point z_0 lies.*

It may be pointed out that the circle $\mathfrak{C}$ through z_0, invariant under $\mathfrak{H}$, is uniquely defined except when $\mathfrak{H}$ is an involution, in which case the two orthogonal pencils defined by the fixed points of $\mathfrak{H}$ consist of invariant circles.

Corollary 2. *Let* $\mathfrak{H}$ *be non-integral* ($c \neq 0$) *so that it has finite poles* $Z_\infty = \mathfrak{H}(\infty)$, $z_\infty = \mathfrak{H}^{-1}(\infty)$. *Then the point*

$$Z_\infty^{(s)} = \mathfrak{H}^s(\infty), \tag{10.77}$$

that is, the inverse pole of the iteration $\mathfrak{H}^s$, *moves on the straight line that carries the diagonal* $\langle z_\infty, Z_\infty \rangle$ *of the characteristic parallelogram of* $\mathfrak{H}$. *If* $\mathfrak{H}$ *is proper hyperbolic* $Z_\infty^{(s)}$ *remains always outside the segment* $\langle \gamma_1, \gamma_2 \rangle$. *If* $\mathfrak{H}$ *is elliptic then the power* $\mathfrak{H}^s$ *for which* $Z_\infty^{(s)}$ *is the common point of the two diagonals of the characteristic rhombus* (§ 8, **e**) *is an involution.*

If $\mathfrak{H}$ is loxodromic or improper hyperbolic, but $k \neq -1$, then the point $z = \mathfrak{H}^s(z_0)$ moves on a loxodromic curve that is the image of a logarithmic spiral by a Moebius transformation (cf. § 8, ex. 6).

Examples

1. Let $\mathfrak{H}$ be a non-elliptic Moebius transformation with two different finite fixed points. Show that the attractive fixed point γ_1 of $\mathfrak{H}$ lies nearer than the repulsive fixed point γ_2 to the pole $Z_\infty = a/c$ of the transformation $\mathfrak{H}^{-1}$(cf. 8.71).

2. *Periodic continued fractions.* Consider a sequence of Moebius transformations with the matrices

$$\mathfrak{H}_1 = \begin{pmatrix} 0 & a_1 \\ 1 & b_1 \end{pmatrix}, \quad \mathfrak{H}_2 = \begin{pmatrix} 0 & a_2 \\ 1 & b_2 \end{pmatrix}, \ldots, \qquad \text{all } a_n \neq 0. \tag{10.8}$$

According to (10.4) it defines the sequence of complex numbers

$$z_1 = \frac{a_1}{b_1 + z_0}, \qquad z_2 = \cfrac{a_1}{b_1 + \cfrac{a_2}{b_2 + z_0}}, \qquad z_3 = \cfrac{a_1}{b_1 + \cfrac{a_2}{b_2 + \cfrac{a_3}{b_3 + z_0}}}, \ldots, \tag{10.81}$$

or, making use of an abbreviated notation,

$$z_n = \frac{a_1}{b_1 +} \frac{a_2}{b_2 +} \cdots \frac{a_n}{+ b_n + z_0} \qquad (n = 1, 2, \ldots).$$

For $z_0 = 0$ these z_n are the 'approximants' of the 'infinite continued fraction,' symbolically represented by

$$f_1 = \frac{a_1}{b_1 +} \frac{a_2}{b_2 +} \cdots \tag{10.82}$$

which is said to be convergent if the sequence of the approximants tends to a limit; this limit is then called the value of the continued fraction and also denoted by f_1.

If, from a certain index on, the sequence $\mathfrak{H}_n$ is periodic, we call (10.82) a periodic continued fraction, purely periodic if the period starts with the very first term so that

$$f_1 = \frac{a_1}{b_1+}\,\frac{a_2}{b_2+}\cdots\frac{a_p}{+b_p+}\,\frac{a_1}{b_1+}\,\frac{a_2}{b_2+}\cdots\frac{a_p}{+b_p+}\cdots.$$

In this case the question of the convergence of f_1 can be decided by means of Theorem B. Let

$$\mathfrak{H} = \mathfrak{H}_1\mathfrak{H}_2\ldots\mathfrak{H}_p$$

be the period matrix, γ_1 its attractive, γ_2 its repulsive fixed point (thus excluding that $\mathfrak{H}$ is elliptic); if none of the first p approximants is equal to γ_2, then $f_1 = \gamma_1$.

As fixed points of a Moebius transformation the numbers γ_1 and γ_2 are the roots of a certain quadratic equation. If γ_1 is the attractive fixed point of $\mathfrak{H}$, then γ_2 is the attractive fixed point of $\mathfrak{H}^{-1}$. Hence we may construct a periodic continued fraction f_2 whose value is equal to γ_2.

For this purpose we notice that for any matrix $\mathfrak{H} = \begin{pmatrix} a & b \\ c & d \end{pmatrix}$ with the determinant $|\mathfrak{H}| = \delta$ we have

$$\delta\mathfrak{H}^{-1} = \mathfrak{J}\mathfrak{H}'\mathfrak{J}^{-1} \qquad \text{where } \mathfrak{J} = \begin{pmatrix} 0 & -1 \\ 1 & 0 \end{pmatrix}$$

$$= \mathfrak{J}\mathfrak{H}_p'\mathfrak{H}_{p-1}'\ldots\mathfrak{H}_1'\mathfrak{J}^{-1}.$$

Since

$$\mathfrak{H}_\nu' = \begin{pmatrix} 0 & 1 \\ a_\nu & b_\nu \end{pmatrix} = a_\nu \begin{pmatrix} 0 & \dfrac{1}{a_\nu} \\ 1 & \dfrac{b_\nu}{a_\nu} \end{pmatrix}, \qquad \mathfrak{H}_\nu'(z) = \frac{1}{b_\nu + a_\nu z}, \quad (\nu = 1,\ldots, p),$$

it follows that we have for f_2 the following sequence $z_0^*, z_1^*, z_2^*,\ldots,$ (corresponding to the sequence $z_1, z_2,\ldots,$ for f_1):

$$z_0^* = \mathfrak{J}(z) = -\frac{1}{z}, \qquad z_1^* = -\frac{1}{\mathfrak{H}_p'(z)} = -(b_p + a_p z),$$

$$z_2^* = -\left(b_p + \frac{a_p}{b_{p-1}+a_{p-1}z}\right), \qquad z_3^* = -\left(b_p + \frac{a_p}{b_{p-1}+}\,\frac{a_{p-1}}{b_{p-2}+a_{p-2}z}\right),\ldots,$$

$$z_{p+1}^* = -\left(b_p + \frac{a_p}{b_{p-1}+}\,\frac{a_{p-1}}{b_{p-2}+}\cdots\frac{a_2}{+b_1+}\,\frac{a_1}{b_p+a_p z}\right)\ldots,$$

The values of these z_n^* for $z = 0$ are the approximants of the continued fraction

$$f_2 = -\left(b_p + \frac{a_p}{b_{p-1}} + \cdots + \frac{a_2}{b_1} + \frac{a_1}{b_p} + \frac{a_p}{b_{p-1}} + \cdots + \frac{a_1}{b_d} + \cdots\right) \tag{10.83}$$

which represents the value γ_2 if none of the first p approximants equals γ_1.

From the arithmetical theory of continued fractions it is known that if all a_ν, b_ν are rational numbers, then $f_1 = \gamma_1$ is irrational and therefore not equal to one of the rational approximants of f_1. Thus f_1, f_2 are the two roots of an irreducible quadratic equation with rational coefficients.[12]

3. Show that the continued fraction $f_1 = \frac{1}{b} + \frac{1}{b} + \ldots$ is always convergent if b is real. For complex b it is divergent if and only if the matrix $\begin{pmatrix} 0 & 1 \\ 1 & b \end{pmatrix}$ belongs to an elliptic Moebius transformation, that is, $b = ib_0$ and $-2 < b_0 < 2$.

4. *Proof of Theorem C for $p > 2$.* Let the iteration for a certain point z_0 be defined by (10.5). Further let z be variable and

$$Z^{(1)} = \mathfrak{H}(z), \qquad Z^{(2)} = \mathfrak{H}(Z^{(1)}), \qquad Z^{(3)} = \mathfrak{H}(Z^{(2)}) \ldots.$$

Then, in view of § 6, **d**, Theorem C, the following relations hold:

$$(z, z_0; z_1, z_2) = (Z^{(1)}, z_1; z_2, z_3) = (Z^{(2)}, z_2; z_3, z_4) = \ldots = (Z^{(p)}, z_0; z_1, z_2),$$

whence $Z^{(p)} = z$ for all z. Thus $\mathfrak{H}^p$ is the identity.

5. Making use of the table (10.53) construct examples of Moebius transformations with integral rational coefficients for each of the possible periods 2, 3, 4, 6.

6. Find all possible orders of Moebius transformations $\mathfrak{H} = \begin{pmatrix} a & b \\ c & d \end{pmatrix}$ whose coefficients are rational complex numbers, that is, $a = a_1 + ia_2$, etc., where $a_1, a_2, b_1, \ldots$, are rational numbers.

7. A function $\phi(z)$ is said to be schlicht (also: univalent or simple) in a domain $\mathfrak{D}$ of the z-plane if, for any two different points z_1, z_2 in $\mathfrak{D}$ also $\phi(z_1) \neq \phi(z_2)$. Thus a schlicht function assumes each of its possible values only once and it possesses a unique, one-valued inverse with the same property. It can be shown that for any function $\phi(z)$, regular and schlicht in $\mathfrak{D}$, the differential coefficient $\phi'(z)$ is different from zero for all z in $\mathfrak{D}$.

[12] A continued fraction f_1 is said to be simple if all $a_n = 1$. For simple periodic continued fractions the last result has first been proved by É. Galois [1] in 1828: 'Démonstration d'un théorème sur les fractions continues périodiques'. Cf. also O. Perron [1], chapters II to III. An exposition of the theory of continued fractions consistently using Moebius transformations has been given recently by K. Kolden [1].

The constant k in Schröder's equation (10.64) is equal to $f'(\gamma)$ if γ is a fixed point of the differentiable function $f(z)$. If $f(z) = \mathfrak{H}(z)$ then

$$k = \frac{ad-bc}{(c\gamma+d)^2}.$$

In the case of Abel's equation (10.67) $f'(\gamma) = 1$. At any rate $\mathfrak{H}'(\gamma)$ represents a value of the characteristic constant of $\mathfrak{H}$ and if $\mathfrak{H}$ has two different finite fixed points γ_1, γ_2, then

$$\mathfrak{H}'(\gamma_1)\mathfrak{H}'(\gamma_2) = 1 \qquad \text{(cf. § 8, \textbf{b})}.$$

8. For the singular matrices $\mathfrak{L}_1$, $\mathfrak{L}_2$ introduced in § 10, **f**, **1** the following relations are valid

$$\mathfrak{L}_1-\mathfrak{L}_2 = (\gamma_2-\gamma_1)\mathfrak{E}, \qquad \mathfrak{L}_1^2 = (\gamma_2-\gamma_1)\mathfrak{L}_1, \qquad \mathfrak{L}_2^2 = (\gamma_1-\gamma_2)\mathfrak{L}_2,$$

$$\mathfrak{L}_1\mathfrak{L}_2 = \mathfrak{L}_2\mathfrak{L}_1 = 0 \qquad \text{(zero matrix)}.$$

The matrix $\mathfrak{L}$ occurring in (10.73) is nilpotent: $\mathfrak{L}^2 = 0$. Hence the matrix $\mathfrak{H}$ of a parabolic Moebius transformation is equal to the sum of a multiple of the unit matrix $\mathfrak{E}$ and a nilpotent matrix. Show that the converse of this statement is also true.

9. To complete the discussion of § 10, **f**, **1**, show that if $\gamma_2 = \infty$ ($\mathfrak{H}(z) = az+b$) then

$$\mathfrak{H}^s = a^s\begin{pmatrix}1 & -\gamma_1\\ 0 & 0\end{pmatrix}+\begin{pmatrix}0 & \gamma_1\\ 0 & 1\end{pmatrix}.$$

10. *Iteration of a Moebius transformation with circle matrix.* Making use of the notations of § 9, ex. 7, we can form the continuous iteration $\mathfrak{K}^s$ according to (10.72) or (10.73) respectively. By means of the relations of ex. 8 it is then readily seen that, together with $\mathfrak{K}$, $\mathfrak{K}^s$ is a normalized circle matrix for all real s, by showing that $\mathfrak{K}^s\overline{\mathfrak{K}^s}$ is a multiple of the unit matrix $\mathfrak{E}$. Therefore the iteration $\mathfrak{K}^s$ is associated with a family of circles which for $s = 1$ contains the circle $\mathfrak{C}$ associated with $\mathfrak{K}$. In the hyperbolic case we have $\gamma_2 = \bar{\gamma}_1$, hence $\mathfrak{L}_2 = \overline{\mathfrak{L}}_1$. These matrices $\mathfrak{L}_1$, $\mathfrak{L}_2$ correspond to the two point circles of the hyperbolic pencil of circles which are interchanged by $\mathfrak{K}$; the real axis is an element of this pencil.

Corresponding results are obtained in the other cases. If $\mathfrak{K}$ is elliptic (and thus its fixed points γ_1, γ_2 real), then the pencil of circles associated with the continuous iteration $\mathfrak{K}^s$ coincides with the elliptic pencil of all circles through γ_1, γ_2, which are interchanged by $\mathfrak{K}$. Finally if $\mathfrak{K}$ is parabolic and γ its fixed point, then the pencil of circles represented by the continuous iteration $\mathfrak{K}^s$ coincides with the parabolic pencil of circles touching the real axis at γ and this is again the pencil of circles interchanged by $\mathfrak{K}$.

Thus the Moebius transformations with circle matrices appear to be of interest, because in this case the matrix of the iteration is representative of the circles of the associated pencil of interchanged circles.

11. Let $\mathfrak{H}$ be elliptic and z_0 be a point on the straight line through the fixed points γ_1, γ_2. Then the equation

$$z = \mathfrak{H}^s(z_0) \tag{10.9}$$

represents the pencil of invariant circles of $\mathfrak{H}$ with z_0 as family parameter. By

$$z_0 = (1-t)\gamma_1 + t\gamma_2$$

one may introduce the real parameter t. If $\mathfrak{H}$ is proper hyperbolic, the family of invariant circles is represented by the same equation (10.9) but with z_0 moving on the line whose points are equidistant from γ_1 and γ_2:

$$z_0 = \tfrac{1}{2}(\gamma_1+\gamma_2)+\tfrac{1}{2}i(\gamma_1-\gamma_2)t.$$

12. If $\mathfrak{H}$ is elliptic, parabolic, or proper hyperbolic, then the family of transformations $\mathfrak{H}^s$ coincides with the group $\mathscr{H}_e$, $\mathscr{H}_p^{(1)}$, $\mathscr{H}_h^+$ of § 9, **f** respectively, that is, the groups containing all Moebius transformations which are products of two inversions within the elliptic, parabolic, hyperbolic pencil (respectively) of real circles interchanged by the transformation $\mathfrak{H}$. With regard to the final Remark of § 9, **f**, it is readily seen that any one of these groups is *cyclic* in the sense that all its elements appear as powers $\mathfrak{H}^s$ with real exponent s. Thus we call $\mathfrak{H}$ a *generator* of the group. Any element different from the identity (unit element) is a generator of the continuous cyclic group. (This is, in general, not so in the case of finite or discrete cyclic groups where all elements are powers with integral exponents of a certain group element.)

13. The similarity invariant $\sigma_n = \sigma(\mathfrak{H}^n)$ for $n = 1, 2, \ldots$, may be written as a polynomial in $\sigma = \sigma_1 = \sigma(\mathfrak{H})$. Observing that the characteristic roots of $\mathfrak{H}^n$ are λ_1^n, λ_2^n if λ_1, λ_2 are those of $\mathfrak{H}$ (cf. § 8, ex. 10), we derive from (8.93)

$$\sigma_n = \frac{(\lambda_1^n+\lambda_2^n)^2}{\lambda_1^n\lambda_2^n}-4 = \frac{\lambda_1^{2n}+\lambda_2^{2n}}{\delta^n}-2$$

$$= \frac{1}{2^{2n}\delta^n}\{[\tau+\sqrt{(\sigma\delta)}]^{2n}+[\tau-\sqrt{(\sigma\delta)}]^{2n}\}-2$$

and by using the relation $\dfrac{\tau}{\sqrt{\delta}} = \sqrt{(\sigma+4)}$ we obtain

$$\sigma_n = \frac{1}{4^n}\{[\sqrt{(\sigma+4)}+\sqrt{\sigma}]^{2n}+[\sqrt{(\sigma+4)}-\sqrt{\sigma}]^{2n}\}-2$$

$$= \frac{1}{2^{2n-1}}\sum_{\nu=0}^{n}\binom{2n}{2\nu}\sigma^\nu(\sigma+4)^{n-\nu}-2.$$

To make the formula valid for $\sigma = 0$ put $0^0 = 1$.[13]

[13] Cf. E. Jacobsthal [3], I.

In particular for $\sigma = -2$ (that is, the invariant of the class whose characteristic parallelogram is a square) one has

$$\sigma_n^{(-2)} = \frac{1}{2^{n-1}} \sum_{\nu=0}^{n} (-1)^\nu \binom{2n}{2\nu} - 2.$$

The numerical values of these sums can be written down immediately. Since the corresponding characteristic constant is $\pm i$, we have by (8.3):

$$\sigma_{2m-1}^{(-2)} = -2, \qquad \sigma_{4m-2}^{(-2)} = -4 \ \text{(involution)}, \qquad \sigma_{4m}^{(-2)} = 0 \ \text{(identity.)}$$

14. The process of the iteration of any non-involutory Moebius transformation $\mathfrak{H}$ can be reduced to a recurrence formula of a certain sequence of polynomials. Let

$$\mathfrak{H} = \begin{pmatrix} a & b \\ c & d \end{pmatrix}, \qquad \tau = a+d \neq 0.$$

Then a Moebius transformation $\mathfrak{T}$ can be found such that

(10.91) $$\mathfrak{T}\mathfrak{H}\mathfrak{T}^{-1} = q\mathfrak{B}, \qquad \mathfrak{B} = \begin{pmatrix} 1 & \beta \\ 1 & 0 \end{pmatrix}.$$

Because

$$\sigma(\mathfrak{H}) = \sigma(\mathfrak{B}) = \frac{\tau^2}{\delta} - 4 = -\frac{1}{\beta} - 4$$

one has

(10.92) $$\beta = -\frac{\delta}{\tau^2} = -\frac{1}{\sigma+4}.$$

If $\mathfrak{H}$ is not an integral Moebius transformation ($c \neq 0$) we may choose an integral Moebius transformation $\mathfrak{T}$ (cf. § 8, **e**, Theorem C), viz.

$$\mathfrak{T} = \begin{pmatrix} c & d \\ 0 & a+d \end{pmatrix},$$

which carries out the similarity transformation (10.91). If $\mathfrak{H}$ is integral,

$\mathfrak{H} = \begin{pmatrix} a & b \\ 0 & 1 \end{pmatrix}$, $a \neq -1$, put

$$\mathfrak{T} = \begin{pmatrix} a & 0 \\ a+1 & -b(a+1) \end{pmatrix}.$$

Since β is defined by, and defines uniquely, the class invariant σ (cf. § 8, **b**, Theorem A), the matrices $\mathfrak{B}$ represent a complete system of non-involutory normal forms. The normal form $\mathfrak{B}$ has the advantage that it can be obtained from $\mathfrak{H}$ by means of a transformation matrix $\mathfrak{T}$ whose elements are very simple rational functions of the elements of $\mathfrak{H}$.

According to Jacobsthal [1] we may now form the iteration of $\mathfrak{B}$ in the following manner:

(10.93) $$\mathfrak{B}^n = \begin{pmatrix} f_n(\beta) & \beta f_{n-1}(\beta) \\ f_{n-1}(\beta) & \beta f_{n-2}(\beta) \end{pmatrix} \qquad (n = 1, 2, \ldots)$$

where $f_n(x)$ is the sequence of the so-called Fibonacci polynomials. It is defined by the recurrence relation

$$f_{n+1}(x) = f_n(x) + x f_{n-1}(x), \qquad f_0(x) = 1, \qquad f_{-1}(x) = 0$$

by means of which (10.93) is easily established by induction. Then

$$\mathfrak{H}^n = q^n \mathfrak{T}^{-1} \mathfrak{B}^n \mathfrak{T}.$$

From the recurrence relation $f_n(x) = \sum_{\nu=0}^{N} \binom{n-\nu}{\nu} x^\nu$, $N = [n/2]$.

Formula (10.93) may be applied to establish the roots and algebraic properties (reducibility or irreducibility) of the Fibonacci polynomials. A Moebius transformation $\mathfrak{H}$, and therefore the corresponding transformation $\mathfrak{B}$ also, has the finite order n if and only if its characteristic constant k is a primitive nth root of unity:

$$k = e^{2\pi i p/n}, \qquad (p, n) = 1.$$

Then $\sigma = k + \bar{k} - 2 = -4(\sin \pi p/n)^2$. With regard to (10.92) put

(10.94) $$\beta_n^{(\nu)} = -\tfrac{1}{4}\frac{1}{(\cos \pi\nu/n)^2}.$$

If $\begin{Bmatrix} n = 2m+1 \\ n = 2m \end{Bmatrix}$ there are $\begin{Bmatrix} [n/2] = m \\ [n/2]-1 = m-1 \end{Bmatrix}$ different values $(\cos \pi\nu/n)^2 \neq 0$.

For all the corresponding $\beta_n^{(\nu)}$ we have $\begin{pmatrix} 1 & \beta_n^{(\nu)} \\ 1 & 0 \end{pmatrix}^n = q_n \mathfrak{E}$. Hence by (10.93)

$$f_{n-1}(\beta_n^{(\nu)}) = 0, \quad \begin{matrix} \nu = 1, \ldots, m & \text{if } n = 2m+1 \\ \nu = 1, \ldots, m-1 & \text{if } n = 2m, \end{matrix}$$

or $$f_{2m}(\beta_{2m+1}^{(\nu)}) = 0, \qquad f_{2m-1}(\beta_{2m}^{(\nu)}) = 0,$$

and since $f_{2m}(x)$ has the degree m we know all its m roots $\beta_{2m+1}^{(\nu)}$; similarly $\beta_{2m}^{(\nu)}$ are the m-1 different roots of $f_{2m-1}(x)$, which has the degree $m-1$.

If l is a divisor of n, then the roots of $f_{l-1}(x)$ are also roots of $f_{n-1}(x)$; thus $f_{l-1}(x)$ is a divisor of $f_{n-1}(x)$ and the polynomial $f_{n-1}(x)$ is reducible if n is not a prime number.

15. If a Moebius transformation $\mathfrak{H} = \begin{pmatrix} a & b \\ c & d \end{pmatrix}$ with integral coefficients and $c \neq 0$ admits an iteration sequence $x_n = \mathfrak{H}^n(x_0)$ of integers $x_0, x_1, x_2, \ldots$, then this sequence is periodic.

Proof. Let $\mathfrak{T} = \begin{pmatrix} c & d \\ 0 & 1 \end{pmatrix}$; then $\mathfrak{T}\mathfrak{H}\mathfrak{T}^{-1}(z^*) = (\tau z^* - \delta)/z^* = \tau - \delta/z^*$ has $x_n^* = cx_n + d = \tau - \delta/x_{n-1}^*$ as integral iteration sequence, which is periodic if and only if x_n is periodic. All x_n^* must be divisors of δ; thus there can be only a finite number of different x_n. If the period has the length $p = 1$, then x_0 is a fixed point of $\mathfrak{H}$. Thus consider $p > 1$. There are no pre-period terms in the sequence; in fact $x_n = x_m$ $(0 < m < n)$ would mean $\mathfrak{H}^n(x_0) = \mathfrak{H}^m(x_0)$ so that x_0 would be a fixed point of $\mathfrak{H}^{n-m}$, and thus of $\mathfrak{H}$.

If the period has length $p > 1$, then $\mathfrak{H}$ is of order p and p is necessarily 2 or 3 or 4 or 6 (cf. Theorems C and D, subsection **c**). Table 10.53 gives the corresponding integral values of σ, and for the existence of an integral iteration sequence of period p one has as necessary condition

$$p = 2:\quad \tau = 0;\quad p = 3:\quad \tau^2 = \delta;\quad p = 4:\quad \tau^2 = 2\delta;\quad p = 6:\quad \tau^2 = 3\delta.$$

For other necessary conditions and a construction of Moebius transformations $\mathfrak{H}$ yielding integral iterations for periods 2, 3, 4, 6 see the paper by Adelman [1].

§ 11. Geometrical characterization of the Moebius transformation

a. *The fundamental theorem.* The Moebius transformation has been introduced by its algebraic definition (6.1) and in § 6, **d**, it has been shown that the Moebius transformations as well as the anti-homographies are circle-preserving transformations.[14] *Are there other circle-preserving transformations in the completed plane?*

The question is similar and in fact closely related to the corresponding problem in projective geometry. It is almost obvious that every invertible linear homogeneous transformation in three homogeneous coordinates represents a collineation in the projective plane, that is, a uniquely invertible transformation carrying points into points and straight lines into straight lines. *Are there other collineations in the real projective plane?*

To both these questions the answer is no. In the second case the answer is given by one of the fundamental theorems of real plane projective geometry.[15] In the first case the answer is the object of the following

Theorem A. *Let $Z = f(z)$ be a schlicht (that is, one-valued and uniquely invertible) mapping of the completed plane into*[16] *itself such that every real circle (or straight line) is mapped onto a real circle or straight line. Then $f(z)$ is either a Moebius transformation or an anti-homography.*

[14] German: Kreisverwandtschaften. Cf. Moebius [1] and [2].

[15] Cf. O. Veblen and J. W. Young [1], vol. I, pp. 188–9. Also H. S. M. Coxeter [2], p. 168; and R. Brauer [1]. See also § 11, **b** and ex. 1.

[16] The word 'into' indicates that the image of the completed plane by the mapping $Z = f(z)$ is assumed to be either the completed plane or a proper part of it. It follows from the theorem that even under this wider assumption the image actually is the completed plane.

Proof. The product of two circle-preserving transformations is again a circle-preserving transformation. All these transformations therefore form a group which contains the group of all Moebius transformations and anti-homographies.

Let $\mathfrak{H}$ be the Moebius transformation which maps $f(0)$, $f(1)$, $f(\infty)$ onto 0, 1, ∞. Then $\mathfrak{H}(f(z))$ and $\overline{\mathfrak{H}(f(z))}$ are also schlicht mappings of the completed plane into itself transforming circles onto circles; they have 0, 1, ∞ as fixed points. One of these mappings —which may be denoted by $g(z)$—maps the point $z = i$ into the upper half plane:

$$\text{Im}\, g(i) > 0. \tag{11.1}$$

Since $g(\infty) = \infty$, straight lines (circles through ∞) are mapped onto straight lines. Since $g(0) = 0$ straight lines through 0 are mapped onto straight lines through 0. In particular the real axis (the line through the fixed points) will be mapped onto itself by the transformation $g(z)$.

Circles that touch one another have exactly one point in common; hence their images also touch one another. In particular parallel straight lines (circles 'touching one another at ∞') are mapped onto parallel straight lines. (Note that parallel straight lines in the completed plane correspond stereographically to circles on the sphere touching one another at the point S, the centre of the projection.)

Now suppose the three points z_0, z_1, z_2 to be non-collinear. Then the straight lines through z_1 parallel to the vector $\overrightarrow{z_0 z_2}$, and through z_2 parallel to the vector $\overrightarrow{z_0 z_1}$, intersect at $z_1 + z_2 - z_0$. Their images are the straight lines through $g(z_1)$ parallel to the vector $\overrightarrow{g(z_0)\, g(z_2)}$, and through $g(z_2)$ parallel to the vector $\overrightarrow{g(z_0)\, g(z_1)}$. Their intersection $g(z_1)+g(z_2)-g(z_0)$ is the image of $z_1+z_2-z_0$; hence

$$g(z_1+z_2-z_0) = g(z_1)+g(z_2)-g(z_0). \tag{11.11}$$

In particular

$$g(z_1 \pm z_2) = g(z_1) \pm g(z_2) \tag{11.12}$$

if 0, z_1, z_2 are not collinear. By (11.12)

$$g(z_1)-g(z) = g(z_1-z) = g[z_1+(-z)] = g(z_1)+g(-z),$$

or,

$$g(-z) = -g(z). \tag{11.13}$$

Now let $0, z_1, z_2$ be collinear; then $z_1 - iz_1, z_2 \pm iz_1$ are not on one line together with 0 except when $z_2 = \mp z_1$, the one case being trivial, the other one covered by (11.13). Hence (11.12) may be applied to $z_1 - iz_1$ and $z_2 \pm iz_1$ in place of z_1 and z_2:

$$\begin{aligned} g(z_1 \pm z_2) &= g[(z_1 - iz_1) \pm (z_2 \pm iz_1)] \\ &= g(z_1 - iz_1) \pm g(z_2 \pm iz_1) \\ &= g(z_1) - g(iz_1) \pm [g(z_2) \pm g(iz_1)] \\ &= g(z_1) \pm g(z_2). \end{aligned}$$

Thus (11.12) is correct for all z_1, z_2.

From (11.12) and (11.13) we conclude by induction that

$$g(nz) = ng(z) \qquad (n = 0, \pm 1, \pm 2, \ldots).$$

Let m be any positive integer. Then

$$g(z) = g\left(m\frac{z}{m}\right) = mg\left(\frac{z}{m}\right) \quad \text{or} \quad g\left(\frac{z}{m}\right) = \frac{1}{m}g(z).$$

Hence

$$g\left(\frac{n}{m}z\right) = g\left(n\frac{z}{m}\right) = ng\left(\frac{z}{m}\right) = \frac{n}{m}g(z)$$

and therefore

$$g(rz) = rg(z) \tag{11.14}$$

for every rational number r. In particular

$$g(r) = rg(1) = r.$$

Thus all rational r are fixed points.

The following geometrical argument shows that the transformation $Z = g(z)$ *preserves orthogonality*. Let $\mathfrak{l}_1, \mathfrak{l}_2$ be two perpendicular straight lines. Draw $\mathfrak{l}_1' \| \mathfrak{l}_1$ and $\mathfrak{l}_2' \| \mathfrak{l}_2$ ($\mathfrak{l}_1' \neq \mathfrak{l}_1$, $\mathfrak{l}_2' \neq \mathfrak{l}_2$). The four lines intersect at the vertices of a rectangle. The intersections of their images are therefore the vertices of a parallelogram. Since the four vertices of the rectangle lie on a circle, the same holds true for their images. Hence the image parallelogram also is a rectangle; thus $g(\mathfrak{l}_1)$ is perpendicular to $g(\mathfrak{l}_2)$. (Cf. § 8, ex. 5 for the notation.)

Now it follows that the imaginary axis is mapped onto itself. In particular

$$g(i) = \lambda i,$$

and by (11.12) we conclude

$$g(\pm 1 + i) = \pm 1 + \lambda i$$

where λ is real. Further, observe that the straight lines through 0 and $\pm 1+i$ are orthogonal; the same must hold for their images passing through the points $\pm 1+\lambda i$ respectively. Therefore $\lambda = \pm 1$, and by (11.1):

$$\lambda = +1.$$

Thus $g(i) = i$ and by (11.14)

$$g(r+si) = r+si \tag{11.15}$$

for all rational r and s.

Now let α be irrational. Let r_0, r_1, r_2 be rational where

$$r_0 < r_1 < \alpha < r_2. \tag{11.16}$$

Let $\mathfrak{C}$ be the circle through r_1, r_2, and i. Every circle through r_0 and α will intersect $\mathfrak{C}$. Hence every circle through $g(\alpha)$ and $g(r_0) = r_0$ will intersect the image circle $g(\mathfrak{C})$, which coincides with $\mathfrak{C}$. As $g(\alpha)$ is real, this implies

$$r_1 < g(\alpha) < r_2.$$

This is true for every pair of rational numbers r_1, r_2 that satisfies (11.16). Thus $g(\alpha) = \alpha$. Similarly $g(\alpha i) = \alpha i$ and

$$g(z) = g(x+yi) = g(x)+g(yi) = x+yi = z$$

for all complex z. Thus $g(z)$ is the identity and

$$f(z) = \mathfrak{H}^{-1}(z) \qquad \text{or} \qquad f(z) = \overline{\mathfrak{H}^{-1}(z)}.$$[17]

Corollary. *A circle-preserving transformation of the completed plane, which preserves the sense of rotation (orientation) at one point, is necessarily a Moebius transformation.*

b. *Complex projective transformations.* The fundamental theorem of plane projective geometry as quoted at the beginning of § 11, **a** gives an algebraic characterization of the collineations in the plane. It can readily be extended to the three-dimensional space and to all spaces of dimension $n > 2$.[18] For $n = 1$, however, that is, for the real projective straight line, the geometrical characterization of a projectivity as a 'collineation' is naturally void. In this case a definition of a projectivity has been introduced in § 7, **b**, and it was shown that every one-dimensional projectivity can be represented by a Moebius

[17] Direct proofs of this Theorem A seem to be so rare in the mathematical literature that C. Caratheodory [2], as late as 1937, found it necessary to publish one. The present proof owes simplifications to suggestions by P. Scherk. Other proofs will be indicated in § 11, **c** and in ex. 1. Cf. also the paper by G. Darboux [1]. A proof by Lie and Scheffers [1], pp. 414-16, takes for granted the most difficult part.

[18] Cf. the paper by R. Brauer [1].

transformation. This result led us immediately to the fundamental theorem of projective geometry of the real straight line (§ 7, Theorem B). If the line is the real axis, then the Moebius transformation representing a projectivity will be real, (cf. the final Remark in § 8, **d**).

The definition of a real one-dimensional projectivity as given in § 7, **b** employs the two-dimensional construction of a perspectivity. We shall now replace this definition by the following 'purely one-dimensional' definition. A projectivity on the x-axis is a one–one correspondence $X = f(x)$ between real points x and X which maps every harmonic set of four points x_1, x_2, x_3, x_4 onto a harmonic set X_1, X_2, X_3, X_4, where $X_\nu = f(x_\nu)$ [cf. (5.61)].

Evidently every real Moebius transformation represents a projectivity in the sense of this definition. A second fundamental theorem of one-dimensional projective geometry states that every projectivity in this sense can be represented by a real Moebius transformation (and is therefore a projectivity in the sense of the former definition. The proof follows readily from the subsequent discussion in conjunction with some remarks made in ex. 2.

Let us define a *projectivity on the complex straight line* (that is, the completed z-plane) as a (not necessarily schlicht) mapping of the completed plane into itself such that every harmonic set $z_1, z_2; z_3, z_4$ is mapped onto a harmonic set $Z_1, Z_2; Z_3, Z_4$. By § 6, Theorem C and ex. 12 it is clear that Moebius transformations and anti-homographies are projectivities on the complex line. The question arises as to whether there are other complex projectivities.

Let $Z = f(z)$ be any complex projectivity. As in the beginning of the proof of Theorem A it is then possible to determine a Moebius transformation $\mathfrak{H}$ such that the complex projectivity

$$g(z) = \mathfrak{H}[f(z)] \tag{11.2}$$

has 0, 1, ∞ as fixed points.

Lemma. *A (real or complex) projectivity* $Z = g(z)$ *which has* 0, 1, ∞ *as fixed points represents an automorphism in the field of the complex numbers, that is*

$$g(z_1+z_2) = g(z_1)+g(z_2) \tag{11.21}$$

$$g(z_1 z_2) = g(z_1)g(z_2) \tag{11.22}$$

for any two complex numbers z_1, z_2.

Proof. For any two different points z_1, z_2:

$$(z_1, z_2; \tfrac{1}{2}(z_1+z_2), \infty) = -1.$$

Thus also

$$(g(z_1), g(z_2); \tfrac{1}{2}[g(z_1)+g(z_2)], \infty) = -1.$$

Since g preserves the harmonic relation, it follows from the same formula that

$$(g(z_1), g(z_2); g[\tfrac{1}{2}(z_1+z_2)], \infty) = -1.$$

Hence

$$g[\tfrac{1}{2}(z_1+z_2)] = \tfrac{1}{2}[g(z_1)+g(z_2)].$$

Put $z_1 = z$ and $z_2 = 0$; then $g(\tfrac{1}{2}z) = \tfrac{1}{2}g(z)$. This means that in the preceding formula the factor $\tfrac{1}{2}$ on both sides can be dropped. Thus (11.21) is proved.

Further, let $z_2 = -z_1 = -z$; then

$$g(z_1+z_2) = g(0) = 0 = g(z_1)+g(z_2) \quad \text{or} \quad g(-z) = -g(z).$$

Making use of the argument which led to the formula (11.14) we obtain from (11.21) immediately

$$g(rz) = rg(z) \qquad (r \text{ rational}), \tag{11.23}$$

which, for $z = 1$, expresses the fact that *all rational points of the real axis are fixed points of the projectivity g.*

In order to prove (11.22) we show first that

$$g(z^2) = [g(z)]^2. \tag{11.24}$$

For this purpose we note that

$$(z, -z; 1, z^2) = -1 \tag{11.25}$$

for all $z \neq 1$. Replacing z by $g(z)$ we obtain

$$(g(z), -g(z); 1, g(z)^2) = -1.$$

On the other hand, since g preserves the harmonic relation, it follows from (11.25)

$$(g(z), -g(z); 1, g(z^2)) = -1.$$

By comparison of the last two relations we obtain (11.24).

Now put $z = z_1+z_2$ in (11.24); then

$$\begin{aligned} g(z^2) &= g(z_1^2)+2g(z_1z_2)+g(z_2^2) \\ &= g(z)^2 = [g(z_1)+g(z_2)]^2 = g(z_1)^2+2g(z_1)g(z_2)+g(z_2)^2 \end{aligned}$$

whence by comparing the middle terms we get (11.22).

Theorem B. *If a complex projectivity in the z-plane either maps into itself at least one real circle, or is continuous on a real circle, then it is a Moebius transformation or an anti-homography.*

Proof. (i) Let $\bar{f}$ be a projectivity mapping into itself the real circle $\mathfrak{C}$. Let $\mathfrak{T}$ be a Moebius transformation such that $\mathfrak{T}(\mathfrak{C})$ is the real axis. Then $f = \mathfrak{T}\bar{f}\mathfrak{T}^{-1}$

is a projectivity mapping the real axis into itself. We determine the corresponding projectivity $g = \mathfrak{H}f$ that has 0, 1, ∞ as fixed points. By our assumption all the values $g(x)$ are real for real x.

First we show that the function $g(x)$ is monotonically non-decreasing. Let x_1, x_2 be two real numbers, $x_2 > x_1$ and $x_2 - x_1 = x^2 > 0$. Then by (11.24)

$$g(x_2) - g(x_1) = g(x_2 - x_1) = g(x^2) = g(x)^2 \geqslant 0.$$

If now α is irrational and r_1, r_2 are two rational numbers such that $r_1 < \alpha < r_2$, then

$$g(r_1) = r_1 < g(\alpha) < g(r_2) = r_2.$$

Hence the real numbers α and $g(\alpha)$ are defined by the same Dedekind section in the set of all rational numbers; they are therefore equal. Thus

$$(11.3) \qquad g(x) = x \quad \text{for all real } x.$$

(ii) The same relation is true if $\tilde{f}(z)$ is assumed to be continuous on a circle $\mathfrak{C}$. Then the corresponding automorphism g is continuous on the real axis. By (11.23) we have $g(r) = r$ for all rational r. By a well-known argument (11.3) again follows.[19]

To complete the proof of the theorem we observe that by (11.3), (11.21), and (11.22)

$$(11.31) \qquad g(x+iy) = g(x)+g(iy) = x+g(i)y$$

for real x and y. Now

$$(1, -1; i, -i) = -1,$$

and therefore

$$\big(1, -1; g(i), -g(i)\big) = -1.$$

The last equation is readily seen to be equivalent to

$$g(i)^2 = -1.$$

Thus there are only the two possibilities

$$g(i) = +i \qquad \text{and} \qquad g(i) = -i,$$

whence by (11.31)

$$(11.32) \qquad g(z) = z \qquad \text{or} \qquad g(z) = \bar{z}.$$

In the first case $\tilde{f}$ is necessarily a Moebius transformation, in the second case an anti-homography.

Remark. Theorem B answers the question for all complex projectivities

[19] Since every real number can be approximated by a sequence of rational numbers, it follows that the values of a continuous function $\phi(x)$ are uniquely defined for all real x, if its values at all rational points $x = r$ only are prescribed.

subject to certain restrictions. However, according to the lemma, *every automorphism g* of the field of all complex numbers, combined with a Moebius transformation or an anti-homography, *will produce a complex projectivity* in the same way as the two *continuous automorphisms* (11.32). Now it is a wellknown, but far from trivial, fact that there is an infinity of (highly discontinuous) automorphisms in the field of all complex numbers.[20] Thus there are as many complex projectivities.

c. *Representation in space.* The elementary properties of the stereographic projection guarantee the existence of a one-to-one correspondence between the circle-preserving transformations on the sphere and those in the completed plane. To the product of two of the spherical transformations corresponds the product of the two corresponding circle-preserving transformations in the plane. The two groups of transformations are thus isomorphic.

We determine first an analytic representation of the spherical circle-preserving transformations. A circle on the sphere defines, and is defined by, a plane in space, represented by an equation (3.4). In order to simplify the following discussion we shall introduce homogeneous coordinates x_1, x_2, x_3, x_4 in the ξ, η, ζ-space so that

$$\xi = \frac{x_1}{x_4}, \qquad \eta = \frac{x_2}{x_4}, \qquad \zeta = \frac{x_3}{x_4}.$$

Further, we replace a, b, c, d by u_1, u_2, u_3, u_4; then the plane (3.4) is given by

(11.4) $$u_1x_1+u_2x_2+u_3x_3+u_4x_4 \equiv u'x = 0$$

and the unit sphere (3.1) by

(11.41) $$x_1^2+x_2^2+x_3^2-x_4^2 \equiv x'Lx = 0, \qquad L = \begin{pmatrix} 1 & 0 & 0 & 0 \\ 0 & 1 & 0 & 0 \\ 0 & 0 & 1 & 0 \\ 0 & 0 & 0 & -1 \end{pmatrix}.$$

The plane (11.4) determines on the sphere (11.41)

(i) a real circle if it intersects the sphere so that its pole (cf. § 3, **c**)

$$x = Lu$$

lies outside the sphere: $x'Lx > 0$;

(ii) a point circle if it touches the sphere so that its pole is a point of the sphere: $x'Lx = 0$;

(iii) an imaginary circle if it has no point in common with the sphere, so that its pole x lies inside the sphere: $x'Lx < 0$ (cf. § 3, **b**).

Every circle-preserving transformation of the sphere onto itself will thus be induced by a plane-preserving transformation of the space onto itself which (1) leaves the equation (11.41) invariant [because of (ii)], and (2) does not change the sign of the quadratic form $x'Lx$ [because of (i) and (iii)], in order to make

[20] Cf. B. Segre [1]. For further literature see L.A., pp. 93–4.

sure that real circles are mapped onto real circles, point circles onto point circles, imaginary circles onto imaginary circles.

According to the fundamental theorem of analytical projective geometry, every plane-preserving transformation (collineation) can be represented by a regular linear homogeneous transformation

$$y = Sx \tag{11.42}$$

in the four homogeneous coordinates x_ν with a regular matrix

$$S = \begin{pmatrix} s_{11} & s_{12} & s_{13} & s_{14} \\ s_{21} & s_{22} & s_{23} & s_{24} \\ s_{31} & s_{32} & s_{33} & s_{34} \\ s_{41} & s_{42} & s_{43} & s_{44} \end{pmatrix}$$

with real coefficients $s_{\mu\nu}$. In the present case we have the condition

$$y'Ly = \sigma^2 x'Lx \qquad (\sigma^2 > 0)$$

or in matrix terms

$$S'LS = \sigma^2 L. \tag{11.43}$$

All projective transformations $y = Sx$ whose matrices S satisfy the condition (11.43) form a group $\mathscr{G}$. Each of these transformations induces a circle-preserving transformation of the unit sphere, and every such circle-preserving transformation induces a transformation of planes intersecting (touching or not intersecting) the sphere onto planes intersecting (touching or not intersecting) the sphere. In virtue of the stereographic projection, the group $\mathscr{G}$ is therefore isomorphic to the group $\mathscr{C}$ of all circle-preserving transformations of the completed plane. We shall now prove:

Theorem C. *Every transformation* (11.42) *of* $\mathscr{G}$ *induces uniquely either a Moebius transformation or an anti-homography in the completed z-plane.*

Thus we obtain a second proof of Theorem A, based this time upon the fundamental theorem of three-dimensional projective geometry.[21]

The arrangement made in the second paragraph of the proof of Theorem A (p. 106) may be resumed. The three points 0, 1, ∞ of the completed z-plane have as images on the sphere the points which in homogeneous coordinates are given by the columns

$$x^{(0)} = \begin{pmatrix} 0 \\ 0 \\ 1 \\ 1 \end{pmatrix}, \qquad x^{(1)} = \begin{pmatrix} 1 \\ 0 \\ 0 \\ 1 \end{pmatrix}, \qquad x^{(\infty)} = \begin{pmatrix} 0 \\ 0 \\ -1 \\ 1 \end{pmatrix} \tag{11.44}$$

[21] A less direct proof along these lines is contained in the book by W. Blaschke [1], §§ 7, 8, 20, 49.

respectively. Thus one has to show that if the transformation S satisfies (11.43) and leaves invariant the three points (11.44) then S corresponds under stereographic projection either to the Moebius transformation $Z = z$ or to the antihomography $Z = \bar{z}$.

Indeed the invariance of the points (11.44) leads to the following relations between the real numbers $s_{\mu\nu}$:

$$Sx^{(0)} = \lambda_0 x^{(0)}, \quad \text{that is,} \quad s_{13}+s_{14} = s_{23}+s_{24} = 0 \qquad \text{(i).}$$

$$s_{33}+s_{34} = s_{43}+s_{44} = \lambda_0 \qquad \text{(ii).}$$

$$Sx^{(1)} = \lambda_1 x^{(1)}, \quad \text{that is,} \quad s_{21}+s_{24} = s_{31}+s_{34} = 0 \qquad \text{(i)}'.$$

$$s_{11}+s_{14} = s_{41}+s_{44} = \lambda_1 \qquad \text{(ii)}'.$$

$$Sx^{(\infty)} = \lambda_\infty x^{(\infty)}, \quad \text{that is,} \quad s_{14}-s_{13} = s_{24}-s_{23} = 0 \qquad \text{(i)}''.$$

$$s_{33}-s_{34} = s_{44}-s_{43} = \lambda_\infty \qquad \text{(ii)}''.$$

By (i) and (i)″, $s_{13} = s_{14} = s_{23} = s_{24} = 0$; by (i)′, $s_{21} = 0$; by (ii)′, $s_{11} = \lambda_1$; by (ii) and (ii)″, $s_{33} = s_{44} = \frac{1}{2}(\lambda_0+\lambda_\infty)$, $s_{34} = s_{43} = -s_{31} = \frac{1}{2}(\lambda_0-\lambda_\infty)$; by (ii)′, $s_{41} = \lambda_1-\frac{1}{2}(\lambda_0+\lambda_\infty)$.

Further, we have the condition (11.43). By working out only the first row of the matrix $S'LS$ one can see that $\lambda_0=\lambda_1=\lambda_\infty=\sigma$ and also $s_{12}=s_{32}=s_{42}=0$. Moreover $s_{22}^2 = \lambda_0^2$; hence

$$S = \begin{pmatrix} \lambda_0 & 0 & 0 & 0 \\ 0 & \pm\lambda_0 & 0 & 0 \\ 0 & 0 & \lambda_0 & 0 \\ 0 & 0 & 0 & \lambda_0 \end{pmatrix}, \tag{11.45}$$

yielding the corresponding transformation

$$y_1 = \lambda_0 x_1, \qquad y_2 = \pm\lambda_0 x_2, \qquad y_3 = \lambda_0 x_3, \qquad y_4 = \lambda_0 x_4.$$

Now, in homogeneous coordinates, the stereographic projection (3.31) is represented by

$$z = \frac{x_1+ix_2}{x_3+x_4}, \qquad Z = \frac{y_1+iy_2}{y_3+y_4}.$$

Thus, from the preceding formulae,

$$Z = \frac{x_1 \pm ix_2}{x_3+x_4} = \begin{cases} z \\ \bar{z} \end{cases},$$

which completes the proof.

Taking the determinant on both sides of (11.43) we have

$$|S|^2 = \sigma^8 \qquad \text{or} \qquad |S| = \pm\sigma^4.$$

Since S and σS represent the same projective transformation (real $\sigma \neq 0$) we may restrict ourselves to those S that satisfy the condition

$$S'LS = L \tag{11.46}$$

so that $|S| = \pm 1$. Under this normalization only the two matrices S and $-S$ represent one and the same element of $\mathscr{G}$. Obviously S and $-S$ have the same determinant.

All transformations in $\mathscr{G}$ whose matrices S have positive determinant form a subgroup $\mathscr{G}_+$ of index 2 in $\mathscr{G}$ (cf. ex. 7).

Theorem D. *In the isomorphism of the group $\mathscr{G}$ with the group $\mathscr{C}$ of all circle-preserving transformations of the completed plane, the group $\mathscr{G}_+$ is isomorphic to the group of all Moebius transformations; to the elements of $\mathscr{G}$ with negative determinants correspond the anti-homographies in $\mathscr{C}$.*

Proof. It has been shown that every Moebius transformation can be represented as a product of two or four inversions. These are elements of $\mathscr{C}$. Thus it will be sufficient to show that the matrices S of the space transformations corresponding to inversions have determinants $|S| < 0$.

Now all inversions are contained in two similarity classes. The inversions of the first kind are all similar to $Z = \bar{z}$. It has been established in the preceding proof that the corresponding space transformation has negative determinant.

All inversions of the second kind (cf. § 9, **d**) are similar to the inversion with respect to the imaginary unit circle $z\bar{z}+1 = 0$, that is

$$Z = -\frac{1}{\bar{z}}$$

or in homogeneous coordinates, using (11.41),

$$\frac{y_1+iy_2}{y_3+y_4} = -\frac{x_3+x_4}{x_1-ix_2} = -\frac{(x_3+x_4)(x_1+ix_2)}{x_1^2+x_2^2} = \frac{x_1+ix_2}{x_3-x_4};$$

this will be satisfied by $y_1 = x_1$, $y_2 = x_2$, $y_3 = x_3$, $y_4 = -x_4$, and the determinant of this transformation has the value -1.

Examples

1. Another proof of Theorem A may be based on the fundamental theorem of real plane projective geometry, making use only of the first part of the proof given in § 11, **a**, viz. up to the invariance of the unit circle. Since the transformation $Z = g(z)$ carries lines into lines, it represents a collineation in the plane, in fact in the sense of projective geometry where the plane is completed not by a single point, but by a straight line at infinity. Because the decision on the nature of

this collineation will be made in the neighbourhood of the point 0, we may ignore the character of the infinity. As our collineation leaves the point 0 invariant, it can be written in the form

$$X = ax+by, \qquad Y = cx+dy, \qquad (ad-bc \neq 0).$$

Because it leaves the x-axis invariant, we have $b = 0$. It carries circles about 0 onto circles about 0; hence $c = 0$ and $a^2 = d^2$, that is, $d = \pm a$. Finally it leaves the point $x = 1$, $y = 0$ invariant; thus $a = 1$. It follows that $X = x$, $Y = \pm y$, q.e.d.

2. A real one-dimensional projectivity, defined as a mapping of the real axis onto itself which maps harmonic sets onto harmonic sets (cf. § 11, **b**), can always be represented by a real Moebius transformation. A proof of this fundamental theorem follows from the discussion in § 11, **b**. In fact the lemma remains true if for the 'field of the complex numbers' we substitute any field of numbers (completed by an element at infinity,[22] in particular the field of all real numbers. It can be shown that the only existing automorphism in the field of all real numbers is the identity (cf. L.A. pp. 92–3). In fact the proof is easily completed from arguments used in § 11, **a**.

3. Making use of the plane stereographic projection (cf. § 7, ex. 8) it can be shown that every projective transformation of the plane which maps the unit circle onto itself corresponds to a mapping of the real axis onto itself and this mapping is represented by a real Moebius transformation.

4. Find all projective transformations in space which leave invariant the unit sphere and have the south pole $x^{(\infty)}$ as fixed point. Derive a representation of the corresponding (integral) Moebius transformation or anti-homography with its coefficients expressed explicitly in terms of the coefficients of the corresponding projective transformation.

5. The projective transformation S which corresponds to the inversion with respect to a given circle $\mathfrak{C}$ may be obtained by means of the Theorem D of § 3. Let a, b, c, d be the coefficients of the equation of the corresponding plane as given by (3.41); then $a, b, c, -d$ are the coordinates of the pole **C** of this plane with respect to the unit sphere. Let x be the column of the homogeneous coordinates $x_1, \ldots, x_4$ of a point **P** on the sphere ($x'Lx = 0$). The coordinates of the inverse point **P*** are then given by

$$x^* = \lambda_1 \begin{pmatrix} a \\ b \\ c \\ -d \end{pmatrix} + \lambda_2 x$$

[22] The proof of the lemma given above is valid for any field $\mathscr{F}$ (in the sense of abstract algebra) which is not of characteristic 2. A. J. Hoffman [1] has shown that the lemma is valid also in fields of characteristic 2.

where λ_1, λ_2 are defined by the condition $x^{*\prime}Lx^* = 0$. Thus $x^* = Sx$, where

$$S = \begin{pmatrix} a^2-b^2-c^2+d^2 & 2ab & 2ac & 2ad \\ 2ab & -a^2+b^2-c^2+d^2 & 2bc & 2bd \\ 2ac & 2bc & -a^2-b^2+c^2+d^2 & 2cd \\ -2ad & -2bd & -2cd & -a^2-b^2-c^2-d^2 \end{pmatrix}.$$

Since all Moebius transformations and anti-homographies can be built up as products of inversions, it follows that all projective transformations leaving invariant the unit sphere can be represented as products of (not more than four) matrices of the type of S.

6. Any projective transformation of the unit sphere onto itself which has the centre of the sphere as fixed point represents a rotation of the sphere about its centre **O** in the case of positive determinant, or a rotation combined with a reflexion in the case of negative determinant. We have the condition

$$S\begin{pmatrix}0\\0\\0\\1\end{pmatrix} = \begin{pmatrix}s_{14}\\s_{24}\\s_{34}\\s_{44}\end{pmatrix} = \begin{pmatrix}0\\0\\0\\\lambda\end{pmatrix}.$$

Since $\lambda \neq 0$ we may put $s_{44} = \lambda = 1$. Thus

$$S = \begin{pmatrix} S_3 & 0 \\ t' & 1 \end{pmatrix}$$

where S_3 is a three-rowed matrix and $t' = (s_{41}, s_{42}, s_{43})$. Now

$$S'LS = \begin{pmatrix} S_3'S_3 - tt' & -t \\ -t' & -1 \end{pmatrix}.$$

Thus by (11.43)

$$t = 0, \qquad S_3'S_3 = E_3$$

where E_3 is the three-rowed unit matrix. Thus S_3 is orthogonal. The corresponding Moebius transformations will be determined in § 12, **b**.

7. The group $\mathscr{L}$ of all matrices S satisfying (11.46) is the so-called *Lorentz group*. The group $\mathscr{G}$ is isomorphic to the quotient group $\mathscr{L}/\mathscr{E}$ where $\mathscr{E}$ is the subgroup of $\mathscr{L}$ consisting of the two matrices E and $-E$. Let

$$S = \begin{pmatrix} S_3 & s \\ t' & \sigma \end{pmatrix}, \qquad s = \begin{pmatrix}s_{14}\\s_{24}\\s_{34}\end{pmatrix}, \qquad t = \begin{pmatrix}s_{41}\\s_{42}\\s_{43}\end{pmatrix}$$

be any element of $\mathscr{L}$. Then by (11.46)

(11.5) $\qquad S_3'S_3 = E_3 + tt', \qquad$ (11.51) $\quad \sigma^2 = 1 + s's.$

Now, for the column t as a point in three-dimensional space with $t't = \rho^2$, a rotation R_3 about 0 can be found such that

$$R_3 t = \begin{pmatrix} \rho \\ 0 \\ 0 \end{pmatrix}.$$

Thus
$$R_3(E_3 + tt')R_3' = \begin{pmatrix} 1+\rho^2 & 0 & 0 \\ 0 & 1 & 0 \\ 0 & 0 & 1 \end{pmatrix}.$$

Hence by (11.5) the determinant

$$|S_3'S_3| = |S_3|^2 = 1+\rho^2,$$

and therefore

$$|S_3| \geqslant 1 \quad \text{or} \quad |S_3| \leqslant -1.$$

Also by (11.51)

$$\sigma \geqslant 1 \quad \text{or} \quad \sigma \leqslant -1.$$

This indicates that the group $\mathscr{L}_+$ of all matrices S with positive determinant, the so-called 'proper Lorentz group,' consists of two disconnected parts

$$\mathscr{L}_{++}: \text{All } S \quad \text{where} \quad |S_3| \geqslant 1, \quad \sigma \geqslant 1$$

$$\mathscr{L}_{--}: \text{All } S \quad \text{where} \quad |S_3| \leqslant -1, \quad \sigma \leqslant -1.$$

All transformations $y = Sx$ with S in $\mathscr{L}_{++}$ form the group $\mathscr{G}_+$. Thus $\mathscr{L}_{++}$ is isomorphic to $\mathscr{G}_+$. The group $\mathscr{L}_{++}$ is a normal divisor of index 4 in $\mathscr{L}$ and a subgroup of index 2 in $\mathscr{L}_+$.

Let $\mathscr{L}_{+-}$ be the system of all S where $|S_3| \geqslant 1, \sigma \leqslant -1$ and $\mathscr{L}_{-+}$ the system of all S where $|S_3| \leqslant -1$, $\sigma \geqslant 1$. Then $\mathscr{L}$ has three different subgroups of index 2, viz.

$$\mathscr{L}_{++} + \mathscr{L}_{--}, \qquad \mathscr{L}_{++} + \mathscr{L}_{+-}, \qquad \mathscr{L}_{++} + \mathscr{L}_{-+}.$$

The quotient group $\mathscr{L}/\mathscr{L}_{++}$ is isomorphic to the four group (cf. § 5, **b**).

III

TWO-DIMENSIONAL NON-EUCLIDEAN GEOMETRIES

§ 12. Subgroups of Moebius transformations

A comprehensive treatment of this extensive topic will not be attempted. We shall concentrate on the geometrical properties of those two subgroups which are of fundamental importance for our exposition of the non-euclidean geometries.

a. *The group $\mathcal{U}$ of the unit circle.* This group consists of all Moebius transformations

$$Z = \mathfrak{H}(z) = \frac{az+b}{cz+d}$$

which map the (circumference of the) unit circle onto itself. It is similar to the group of all real Moebius transformations (cf. § 8, **d**). If $\mathfrak{T}$ is a Moebius transformation that maps the real axis onto the unit circle $|z| = 1$ (cf. § 6, ex. 11) then $\mathfrak{T}^{-1}\mathfrak{H}\mathfrak{T}$ represents a real Moebius transformation. Both groups are therefore isomorphic.

The transformation $\mathfrak{H}$ maps every pair of mutually inverse points z, $1/\bar{z}$ onto a pair of mutually inverse points Z, $1/\bar{Z}$. Hence

$$\frac{1}{\bar{Z}} = \frac{b\bar{z}+a}{d\bar{z}+c} \quad \text{or} \quad Z = \frac{\bar{d}z+\bar{c}}{\bar{b}z+\bar{a}}$$

which leads to the matrix condition

$$\begin{pmatrix} a & b \\ c & d \end{pmatrix} = q\begin{pmatrix} \bar{d} & \bar{c} \\ \bar{b} & \bar{a} \end{pmatrix} \qquad (q \neq 0).$$

Assuming that $a \neq 0$ we conclude from $a = q\bar{d} = q\bar{q}a$ that $q\bar{q} = 1$. Let

$$q = e^{2i\phi}.$$

Then $d = e^{2i\phi}\bar{a}$, $c = e^{2i\phi}\bar{b}$. By formally replacing $e^{-i\phi}a$ and $e^{-i\phi}b$ by a and b respectively, we obtain the matrix $\mathfrak{H}$ in the form

$$(12.1) \qquad \mathfrak{H} = \begin{pmatrix} a & b \\ \bar{b} & \bar{a} \end{pmatrix}, \quad Z = \frac{az+b}{\bar{b}z+\bar{a}},$$

which is unique up to a real factor of proportionality. Its determinant and trace

$$(12.11) \qquad \delta = |\mathfrak{H}| = a\bar{a}-b\bar{b}, \quad \tau = a+\bar{a}$$

are then real. If $a = 0$, then $bc \neq 0$ and $b = q\bar{c} = q\bar{q}b$. Thus $|q| = 1$ and (12.1) again follows.

The transformations $\mathfrak{H}$ mapping the interior of the unit circle onto its interior (and therefore the exterior onto the exterior) form a subgroup of index 2 in the group of the unit circle. For each of these it is necessary and sufficient that $\mathfrak{H}(0) = b/\bar{a}$ be an inner point of the unit circle. This is equivalent to $|b| < |a|$ or to

$$\delta > 0. \tag{12.12}$$

We shall denote this subgroup by $\mathscr{U}_+$ and call it the *proper group* of the unit circle.

The improper mappings of the unit circle onto itself are characterized by the condition $\delta < 0$. They interchange the exterior and the interior of the circle.

From the discussion of § 8, **d**, it is evident that the group $\mathscr{U}_+$ consists of elliptic, parabolic, and proper hyperbolic Moebius transformations. Every improper mapping of the unit circle, however, must be improper hyperbolic.

If $a = 0$ the transformation (12.1) will be a hyperbolic involution. If $a \neq 0$, let

$$\mathfrak{H}^{-1}(0) = -\frac{b}{a} = z_1, \qquad \operatorname{arc} a = \alpha;$$

then

$$Z = \mathfrak{H}(z) = \frac{a}{\bar{a}} \frac{z-z_1}{-\bar{z}_1 z+1} = e^{2i\alpha}\mathfrak{H}_{z_1}(z) \tag{12.13}$$

where

$$Z = \mathfrak{H}_{z_1}(z) = \frac{z-z_1}{-\bar{z}_1 z+1}, \qquad \mathfrak{H}_{z_1} = \begin{pmatrix} 1 & -z_1 \\ -\bar{z}_1 & 1 \end{pmatrix} \tag{12.14}$$

is a transformation of the unit circle onto itself which maps z_1 into 0. Since

$$\sigma(\mathfrak{H}_{z_1}) = \frac{4}{1-z_1\bar{z}_1} - 4,$$

the transformation $\mathfrak{H}_{z_1}$ is hyperbolic for every z_1 unless $z_1 = 0$ or $|z_1| = 1$, proper hyperbolic if $|z_1| < 1$, improper if $|z_1| > 1$. Its fixed points are the roots of the quadratic equation $\bar{z}_1 z^2 = z_1$; therefore, if

$$z_1 = r_1 e^{i\theta_1},$$

they are the two diametrically opposite points $\pm e^{i\theta_1}$ of the unit circle. By (8.43) the characteristic constant of $\mathfrak{H}_{z_1}$ is given by

$$k = \frac{1+\bar{z}_1 e^{i\theta_1}}{1-\bar{z}_1 e^{i\theta_1}} = \frac{1+r_1}{1-r_1};$$

thus

$$r_1 = \frac{k-1}{k+1}, \qquad z_1 = \frac{k-1}{k+1} e^{i\theta_1}.$$

Now let $\mathfrak{H}_{z_1}$ be proper hyperbolic (in $\mathscr{U}_+$); thus $0 < r_1 < 1$ $(k > 1)$. Using (10.72) we form the continuous iteration $\mathfrak{H}_{z_1}^s$ of $\mathfrak{H}_{z_1}$ for real exponent s by replacing k wherever it occurs in $\mathfrak{H}_{z_1}$ by k^s. We consider

$$z_s = \frac{k^s - 1}{k^s + 1} e^{i\theta_1}.$$

For $-\infty < s < \infty$ the point z_s ranges through the whole open segment (diameter of the unit circle) whose endpoints are the fixed points $-e^{i\theta_1}$ $(s = -\infty)$ and $e^{i\theta_1}$ $(s = \infty)$ of $\mathfrak{H}$. Then

$$\text{(12.15)} \qquad \mathfrak{H}_{z_1}{}^s(z) = \mathfrak{H}_{z_s}(z), \qquad \mathfrak{H}_{z_s} = \begin{pmatrix} 1 & -z_s \\ -\bar{z}_s & 1 \end{pmatrix}.$$

Subjecting the transformation $\mathfrak{H}_{z_1}{}^s$ to a rotation about 0, we obtain the similar transformation

$$\text{(12.16)} \qquad Z = e^{it}\mathfrak{H}_{z_1}{}^s(e^{-it}z) = \frac{z - e^{it}z_s}{-e^{-it}\,\bar{z}_s z + 1}$$

which is again a hyperbolic transformation, viz. the one which transforms the point $e^{it}z_s$ into 0. By following up the mapping (12.16) with a suitable rotation we may obtain a representation of any given proper transformation of the unit circle. Hence

Theorem A. *Every element of the proper group of the unit circle can be written in the form*

$$\mathfrak{H} = \mathfrak{R}_1{}^{t_1}\mathfrak{H}_{\frac{1}{2}}{}^s\mathfrak{R}_1{}^{t_2} \qquad (s, t_1, t_2 \text{ real}),$$

where $\mathfrak{R}_1 = \begin{pmatrix} e^i & 0 \\ 0 & 1 \end{pmatrix}$ *and* $\mathfrak{H}_{\frac{1}{2}} = \begin{pmatrix} 1 & -\frac{1}{2} \\ -\frac{1}{2} & 1 \end{pmatrix}$, *the matrix of the hyperbolic transformation which maps* $z = \frac{1}{2}$ *onto* 0 *and has* -1 *and* $+1$ *as fixed points.*

Finally we notice a simple consequence of (12.13). Every transformation $\mathfrak{H}$ of the unit circle which has the centre 0 as fixed point ($z_1 = 0$ or $b = 0$) is necessarily a pure rotation about 0.

b. *The group $\mathscr{R}$ of rotational Moebius transformations.* In § 11, ex. 6, it has been shown that all projective transformations of the unit sphere which have the centre **O** as fixed point form a group that is isomorphic to the group of the three-rowed orthogonal matrices S_3. Those with positive determinant represent rotations of the sphere about **O**. They map straight lines through **O** onto lines through **O** and therefore carry pairs of diametrically opposite points on the sphere onto pairs of diametrically opposite points. Stereographically they correspond to Moebius transformations

$$Z = \mathfrak{H}(z) = \frac{az + b}{cz + d}$$

which map pairs of antipodal points z, $-1/\bar{z}$ onto pairs of antipodal points Z, $-1/\bar{Z}$ in the plane [cf. § 3, ex. 1 and 3 (a)]. Thus

$$-\frac{1}{\bar{Z}} = \mathfrak{H}\left(-\frac{1}{\bar{z}}\right) = \frac{b\bar{z}-a}{d\bar{z}-c}$$

or

$$Z = \frac{\bar{d}z-\bar{c}}{-\bar{b}z+\bar{a}}$$

whence

$$\begin{pmatrix} a & b \\ c & d \end{pmatrix} = q\begin{pmatrix} \bar{d} & -\bar{c} \\ -\bar{b} & \bar{a} \end{pmatrix}.$$

As in § 12, **a** it is seen that $q\bar{q} = 1$, $q = e^{2i\phi}$, and thus

$$e^{-i\phi}\begin{pmatrix} a & b \\ c & d \end{pmatrix} = e^{i\phi}\begin{pmatrix} \bar{d} & -\bar{c} \\ -\bar{b} & \bar{a} \end{pmatrix}.$$

Hence if we replace $e^{-i\phi}a$ by a, etc., it follows that

$$\begin{pmatrix} a & b \\ c & d \end{pmatrix} = \begin{pmatrix} \bar{d} & -\bar{c} \\ -\bar{b} & \bar{a} \end{pmatrix}$$

or

(12.2) $$\mathfrak{H} = \begin{pmatrix} a & b \\ -\bar{b} & \bar{a} \end{pmatrix}, \quad \text{and} \quad Z = \frac{az+b}{-\bar{b}z+\bar{a}}$$

expresses a rotational Moebius transformation.

Any such transformation $\mathfrak{H}$ is seen to be elliptic; for

$$\delta = a\bar{a}+b\bar{b} > 0, \quad \tau = a+\bar{a},$$

and therefore

$$\frac{\tau^2}{\delta} = \frac{(a+\bar{a})^2}{a\bar{a}+b\bar{b}} \leqslant 4\left(\frac{\text{Re } a}{|a|}\right)^2 \leqslant 4$$

so that

$$-4 \leqslant \sigma(\mathfrak{H}) \leqslant 0,$$

while $\sigma(\mathfrak{H}) = 0$ implies that a is real, $b = 0$, and thus $\mathfrak{H} = q\mathfrak{E}$.

Again we note that those rotational Moebius transformations which have $z = 0$ as fixed point are pure rotations about 0.

The matrix (12.2) is defined up to an arbitrary real factor. Thus $\mathfrak{H}$ may be further normalized by requiring that

(12.21) $$\delta = |\mathfrak{H}| = |a|^2+|b|^2 = 1.$$

Then the matrix $\mathfrak{H}$ is seen to be *unitary*, that is,

(12.22) $$\bar{\mathfrak{H}}'\mathfrak{H} = \mathfrak{E} \quad \text{or} \quad \mathfrak{H}^{-1} = \bar{\mathfrak{H}}'.$$

Conversely if $\mathfrak{H} = \begin{pmatrix} a & b \\ c & d \end{pmatrix}$ is unitary and $|\mathfrak{H}| = 1$, then

$$\mathfrak{H}^{-1} = \begin{pmatrix} d & -b \\ -c & a \end{pmatrix} = \begin{pmatrix} \bar{a} & \bar{c} \\ \bar{b} & \bar{d} \end{pmatrix}$$

and $\mathfrak{H}$ has indeed the form (12.2).

The group $\mathscr{R}$ of the rotational Moebius transformations could have been introduced as the *group of all those Moebius transformations which leave the imaginary unit circle* $z\bar{z}+1 = 0$ *invariant*, thus showing its analogy with the group $\mathscr{U}_+$ of transformations leaving the real unit circle $z\bar{z}-1 = 0$ invariant. Indeed every transformation leaving invariant the imaginary unit circle maps pairs of points, symmetric with respect to this circle, onto such symmetric pairs; such pairs of points being antipodal, it is a rotational Moebius transformation.

A *similarity normal form* $\mathfrak{H}^* = \mathfrak{T}\mathfrak{H}\mathfrak{T}^{-1}$ of a rotational $\mathfrak{H}$ may be obtained by means of a transformation $\mathfrak{T}$ of the group $\mathscr{R}$. We observe that if γ is one of the fixed points of $\mathfrak{H}$, then the other one is the antipodal point $-1/\bar{\gamma}$. Let **P** and $-$**P** be the corresponding (diametrically opposite) spherical points. By a rotation $\mathfrak{T}$ of the sphere it is possible to turn **P** into **N** [north pole (0, 0, 1)] and $-$**P** into **S** [south pole (0, 0, -1)]. The corresponding Moebius transformation $z^* = \mathfrak{T}(z)$ is rotational and takes $\gamma \to 0$, $-1/\bar{\gamma} \to \infty$. Thus

$$(12.23) \qquad Z^* = e^{i\psi}z^* = \mathfrak{H}^*(z^*), \qquad \mathfrak{H}^* = \begin{pmatrix} e^{i(\psi/2)} & 0 \\ 0 & e^{-i(\psi/2)} \end{pmatrix},$$

is the normal form; the matrix $\mathfrak{H}^*$ is uniquely defined up to its sign (cf. ex. 3).

If $a = 0$ the mapping $\mathfrak{H}$ is an involution $Z = -b/\bar{b}z = -e^{2i\beta}/z$; its fixed points are the antipodals $\pm ie^{i\beta}$.

If $a \neq 0$ we introduce as in § 12, **a** the point

$$z_1 = \mathfrak{H}^{-1}(0) = -\frac{b}{a} = r_1 e^{i\theta_1} \qquad (r_1 > 0)$$

which by the rotational transformation $\mathfrak{H}$ of (12.2) is mapped into 0. Then

$$(12.24) \qquad Z = \mathfrak{H}(z) = \frac{a}{\bar{a}}\frac{z-z_1}{\bar{z}_1 z+1} = e^{2i\alpha}\frac{z-z_1}{\bar{z}_1 z+1} = e^{2i\alpha}\mathfrak{H}_{z_1}(z).$$

The transformation

$$Z = \mathfrak{H}_{z_1}(z), \qquad \mathfrak{H}_{z_1} = \begin{pmatrix} 1 & -z_1 \\ \bar{z}_1 & 1 \end{pmatrix}$$

has as fixed points the two antipodal points $\pm ie^{i\theta_1}$ of the real unit circle. Its characteristic constant

$$k = \frac{1-\bar{z}_1 ie^{i\theta_1}}{1+\bar{z}_1 ie^{i\theta_1}} = \frac{1-ir_1}{1+ir_2} = e^{-i\kappa}$$

is independent of θ_1, and the real constant κ, defined up to an additive integral multiple of 2π, may be taken to be positive and $\leqslant \pi$. Then

$$r_1 = \tan \frac{\kappa}{2}.$$

The continuous iteration $\mathfrak{H}_{z_1}{}^s$ for real s is found by replacing the constant k in $\mathfrak{H}_z$ by k^s, that is, by replacing κ by $s\kappa$, where s ranges through an interval such that $0 \leqslant \kappa s < \pi$; thus

$$(12.25)\quad \mathfrak{H}_{z_1}^s(z) = \mathfrak{H}_{z_s}(z), \qquad \mathfrak{H}\bar{z}_s = \begin{pmatrix} 1 & -z_s \\ \bar{z}_s & 1 \end{pmatrix}, \qquad z_s = \tan\left(\frac{\kappa s}{2}\right)e^{i\theta_1}.$$

If $z_1 = 1$, we have $\mathfrak{H}_1 = \begin{pmatrix} 1 & -1 \\ 1 & 1 \end{pmatrix}$, $\kappa = \pi/2$, $z_s = \tan(\pi s/4)$. Thus all the values which z_s can assume will be assumed for s in the interval $0 \leqslant s < 2$. The limit case $s = 2$ corresponds to the involution ($a = 0$).

By means of the argument applied in the proof of Theorem A we obtain

Theorem B. *Every rotational Moebius transformation can be written in the form*

$$\mathfrak{H} = \mathfrak{R}^{t_1}\mathfrak{H}_1{}^s\,\mathfrak{R}^{t_2} \qquad (0 \leqslant s \leqslant 2)$$

where $\mathfrak{R}$ *is a rotation about* 0 *and* $\mathfrak{H}_1$ *the elliptic transformation which maps* $z = 1$ *into* 0 *and has* $\pm i$ *as its fixed points.*

c. *Normal forms of bundles of circles.* Every pencil of circles in the plane can by a suitable Moebius transformation be transformed onto its normal form (cf. § 4, ex. 2), namely

in the elliptic case:	all straight lines through 0,
in the parabolic case:	all lines parallel to the real axis,
in the hyperbolic case:	all circles about 0.

We shall now establish the corresponding situation for bundles of circles, in which the normal forms are those described in § 4, **b**, and we shall use the same notations as in that section.

(i) Given an elliptic bundle, its image on the sphere is determined by the point **P** through which pass the planes of the corresponding spherical circles. By a suitable rotation of the sphere we take this point **P** into a point of the ζ-axis whose homogeneous coordinates are $(0, 0, \rho, 1)$ $(0 \leqslant \rho = \mathbf{OP} < 1)$. The projective transformation given by the matrix

$$S = \begin{pmatrix} 1 & 0 & 0 & 0 \\ 0 & 1 & 0 & 0 \\ 0 & 0 & \tilde{\rho} & -\rho\tilde{\rho} \\ 0 & 0 & -\rho\tilde{\rho} & \tilde{\rho} \end{pmatrix}, \qquad \tilde{\rho} = \frac{1}{\sqrt{(1-\rho^2)}}$$

maps the unit sphere onto itself and the point $(0, 0, \rho, 1)$ on the origin $\mathbf{O} = (0, 0, 0, 1)$. To the two employed transformations corresponds a Moebius transformation in the plane which maps the given elliptic bundle onto its normal form, the system of all circles orthogonal to the imaginary unit circle.

(ii) A parabolic bundle on the sphere consists of the circles cut out on the sphere by all planes passing through a point **P** on the surface of the sphere. A rotation will take this point into the south pole **S** of the sphere; the bundle will then consist of all circles through **S** whose stereographic images in the plane are the elements of the normal form: all straight lines in the plane.

(iii) A hyperbolic bundle on the sphere consists of the intersections of the sphere with the planes through a point **P** outside the sphere. By a projectivity one can map the sphere onto itself and **P** into the point at infinity of the ζ-axis. The bundle will then consist of the circles on the sphere whose planes are parallel to the ζ-axis. Thus all these circles are orthogonal to the equator. The projectivity corresponds to a Moebius transformation in the plane which transforms the given hyperbolic bundle onto its normal form, that is, the bundle of all the circles orthogonal to the unit circle.

Remark. Every bundle of circles may be defined analytically by a linear homogeneous condition between the coefficients A, B, C, D of the circle equation (1.2). For the three normal forms these conditions are given in § 4, ex. 7. If the Moebius transformation by which a given bundle is turned into its normal form is known, then the corresponding condition for the given bundle in its original position can be derived from these equations by means of (6.42).

d. *The bundle groups.* It has been shown (§ 9, **e**, and § 10, ex. 12) that all the Moebius transformations which are products of inversions within a pencil of circles form the (cyclic) group of all Moebius transformations interchanging the circles of the pencil. The corresponding situation in the case of a bundle will be discussed now.

Every inversion with respect to a circle of a bundle $\mathfrak{b}$ may be called 'an inversion within the bundle $\mathfrak{b}$ (cf. § 9, **e**). The bundle consists of all the circles orthogonal to a given circle $\mathfrak{P}$ (§ 4, **b**, Theorem B). Every inversion within the bundle will therefore map $\mathfrak{P}$ onto itself (cf. § 2, ex. 1, and § 6, ex. 13) and every circle of the bundle onto some circle of the bundle. Thus the product of any two inversions within the bundle will be a Moebius transformation mapping the bundle onto itself.

The converse is not true without restriction:

Theorem C. *Every Moebius transformation* $\mathfrak{H}$ *of the bundle* $\mathfrak{b}$ *onto itself is the product of two inversions within the bundle if and only if* $\mathfrak{b}$ *is elliptic or hyperbolic.*

Proof. If $\mathfrak{b}$ is elliptic or hyperbolic, its circle $\mathfrak{P}$ is imaginary or real, and not a point circle. Therefore the group $\mathscr{G}_\beta$ of all Moebius transformations mapping $\mathfrak{b}$ onto itself contains no loxodromic transformation. Now let $\mathfrak{H}$ be an element of $\mathscr{G}_\beta$. The pencil of all circles interchanged by $\mathfrak{H}$ (orthogonal to the pencil of invariant circles) is contained in $\mathfrak{b}$, and by § 9, **e**, Theorem E, we know that $\mathfrak{H}$ is the product of two inversions within this pencil, and therefore within $\mathfrak{b}$.

If $\mathfrak{b}$ is parabolic, it is readily seen that the group $\mathscr{G}_\beta$ contains loxodromic transformations. Indeed it may be assumed that $\mathfrak{b}$ is the normal form of a parabolic bundle, that is, the system of all straight lines in the plane (the point circle $\mathfrak{P}$ being the point at infinity). The group $\mathscr{G}_\beta$ then is the group of all integral Moebius transformations $Z = az+b$ (cf. § 6, ex. 4, and § 9, ex. 6) which are loxodromic if a is not real and not of modulus one. (Cf. § 9, **e**, Theorem F, Corollary.)

On the other hand, we consider again the inversions within the parabolic bundle; they are symmetries with respect to straight lines (cf. § 2, ex. 8) and thus they preserve all distances in the plane. Hence the Moebius transformations that are products of such inversions also preserve the distances. They are therefore plane euclidean motions or displacements, and can be represented by

$$Z = e^{i\alpha}z+b \qquad \text{(real } \alpha). \tag{12.3}$$

The same follows from the fact that they are elliptic or parabolic integral Moebius transformations. The group (12.3) will be denoted by $\mathscr{E}$. Its elements depend on three independent real parameters: α, b_1, b_2 if $b = b_1+ib_2$. (Cf. ex. 7.)

If $\mathfrak{b}$ is elliptic, then its normal form is the bundle of the stereographic images of all great circles on the sphere. Every projective transformation in space mapping great circles onto great circles maps planes through **O** onto planes through **O** and lines through **O** onto lines through **O**; therefore pairs of diametrically opposite points of the sphere are mapped onto pairs of diametrically opposite points. Thus the transformation is a rotation about **O**. The group of the elliptic bundle in its normal form is therefore the group $\mathscr{R}$ of all rotational Moebius transformations.

Remark. In view of § 3, **c**, Theorem D, we conclude from Theorem C the well-known fact that every space rotation about **O** can be represented as the product of two symmetries with respect to planes through **O**.

If $\mathfrak{b}$ is hyperbolic, then its normal form is the bundle of all circles orthogonal to the real unit circle. It contains elliptic, parabolic, and hyperbolic pencils and thus also imaginary circles. Its group is the group $\mathscr{U}$ of all Moebius transformations mapping the unit circle onto itself. The fundamental circles of the inversions, as whose product an element $\mathfrak{H}$ of $\mathscr{U}$ may be obtained, can be chosen real except when $\mathfrak{H}$ is improper hyperbolic in which case one of these circles must be imaginary.

e. *Transitivity of the bundle groups.* A group $\mathscr{G}$ of transformations $Z = \mathfrak{T}(z)$ of the completed plane onto itself is said to be *transitive* in a domain $\mathscr{D}$ of the plane if for every pair of points z_1, Z_1 in D there is at least one transformation $\mathfrak{T}$ in $\mathscr{G}$ such that $Z_1 = \mathfrak{T}(z_1)$. The domain $\mathscr{D}$ (which may also be the whole plane) is called a *domain of transitivity* of the group $\mathscr{G}$.

The group $\mathscr{G}$ is said to be p-fold transitive in $\mathscr{D}$ if for any two sets of p points: $z_1, z_2,\ldots, z_p$ and $Z_1, Z_2,\ldots, Z_p$ there is at least one $\mathfrak{T}$ in $\mathscr{G}$ such that

$$Z_1 = \mathfrak{T}(z_1), \qquad Z_2 = \mathfrak{T}(z_2),\ldots, Z_p = \mathfrak{T}(z_p).$$

Obviously every p-fold transitive group is also $(p-1)$-fold transitive in the same domain $\mathscr{D}$. If $p = 1$ the group is simply transitive.

From § 6, **d**, Theorem D, it is evident that the group $\mathscr{M}$ of all Moebius transformations is three-fold, but not four-fold transitive in the completed plane. The real axis is a domain of three-fold transitivity for the group of all real Moebius transformations. The group of all integral Moebius transformations is two-fold, but not three-fold transitive in the (open) plane of complex numbers (without the point at infinity).

For the material of chapter III the following fact is of fundamental importance:

Theorem D. *Every group $\mathscr{G}$ of all Moebius transformations obtained as products of inversions within a bundle of circles is simply, but not two-fold, transitive in a domain $\mathscr{D}$. For the group of the elliptic bundle, $\mathscr{D}$ is the completed plane. For the group of the parabolic bundle, $\mathscr{D}$ is a dotted plane (that is, the completed plane from which one point has been removed).*[1] *For the proper group of the hyperbolic bundle, $\mathscr{D}$ is the interior of a circle.*

Proof. It will be sufficient to prove the theorem for the three groups $\mathscr{R}, \mathscr{E}, \mathscr{U}_+$.

(i) In the case of the elliptic bundle it will be sufficient to show that the sphere is a domain of simple transitivity for the group of all rotations about its centre. In fact every point of the sphere can be transported into every other point of the sphere by a suitable rotation. But the rotation preserves the distance between any two points of the sphere; thus a given pair of points cannot be turned onto every other pair of points.

(ii) A similar argument can be applied in the parabolic case. The complex number plane, dotted at infinity, is a domain of simple transitivity for the group $\mathscr{E}$ of euclidean displacements. A pair of points z_1, z_2 can be transported into another pair Z_1, Z_2 if and only if $|Z_1-Z_2| = |z_1-z_2|$. Thus the group is not two-fold transitive.

(iii) The hyperbolic group $\mathscr{U}_+$ has the interior of the unit circle $|z| < 1$ as domain of simple transitivity. In fact, by the transformation (12.14) any given

[1] In the case of the parabolic bundle in its normal form (all lines in the plane), $\mathscr{D}$ is the (open) complex number plane, corresponding to the dotted sphere, that is, the sphere without the south pole **S**.

point z_1 ($|z_1| < 1$) will be mapped into the centre 0, but any subsequent transformation, designed to take a second point into a prescribed position, was shown to be a pure rotation about 0 [cf. (12.13)], thus preserving the distance from 0.

In the first two cases the proof was based on the geometrically evident existence of an invariant distance function, that is, a function $d(z_1, z_2)$ of two points z_1, z_2 that satisfies the following conditions:

(12.4) $$d[\mathfrak{H}(z_1), \mathfrak{H}(z_2)] = d(z_1, z_2)$$

for every element $\mathfrak{H}$ of the group; and

(12.41) $$d(z_1, z_2) \neq 0 \quad \text{if} \quad z_1 \neq z_2; \qquad d(z_1, z_1) = 0.$$

In the third case a two-point invariant is given by the cross ratio

(12.42) $$d_{-1}(z_1, z_2) = \left(z_1, z_2; \frac{1}{\bar{z}_1}, \frac{1}{\bar{z}_2}\right).$$

Indeed, let $Z = \mathfrak{H}(z)$ be an element of the group $\mathscr{U}_+$ (§ 12, **a**). Then $1/\bar{Z} = \mathfrak{H}(1/\bar{z})$, and therefore

$$d_{-1}(Z_1, Z_2) = \left(\mathfrak{H}(z_1), \mathfrak{H}(z_2); \mathfrak{H}\left(\frac{1}{\bar{z}_1}\right), \mathfrak{H}\left(\frac{1}{\bar{z}_2}\right)\right) = d_{-1}(z_1, z_2).$$

Thus in the hyperbolic case the proof can be carried through in the same way as in the two other cases.

Remark. It follows from § 6, **d**, Theorem D, that in the completed plane the group $\mathscr{M}$ of all Moebius transformations is 'exactly three-fold transitive' in the strong sense that there is one, and not more than one, Moebius transformation that maps three given distinct points onto three given distinct points. It is remarkable that every group of continuous transformations of the completed plane onto itself, which is exactly three-fold transitive, is similar to the group $\mathscr{M}$; this means that after a topological transformation of the completed plane onto itself (which may be the identity) the given exactly three-fold transitive group coincides with $\mathscr{M}$. This was shown in 1940 by B. de Kérekjártó [2]. A new approach to the problem has been made recently by J. Tits [1] and [2]. For another, more algebraic characterization of the group $\mathscr{M}$ cf. the paper by F. Bachmann [1] and also his book [2], § 11.

EXAMPLES

1. Find the intersection of the two groups $\mathscr{R}$ and $\mathscr{U}$, that is, the set (subgroup) of all transformations they have in common.

2. For two distinct points z_1, z_2 a rotational Moebius transformation $\mathfrak{R}$ can be found such that $\mathfrak{R}(z_1)$ and $\mathfrak{R}(z_2)$ are symmetric with respect to the real unit circle.

Observe that there is a unique great circle Γ with respect to which the spherical points $\mathbf{P}_1$, $\mathbf{P}_2$, corresponding stereographically to z_1, z_2, are symmetric. The stereographic image $\mathfrak{C} = \begin{pmatrix} A & B \\ \bar{B} & -A \end{pmatrix}$ of Γ in the plane belongs to the hyperbolic pencil whose point circles lie at z_1, z_2, viz.

$$\lambda_1 \begin{pmatrix} 1 & -\bar{z}_1 \\ -z_1 & z_1\bar{z}_1 \end{pmatrix} + \lambda_2 \begin{pmatrix} 1 & -\bar{z}_2 \\ -z_2 & z_2\bar{z}_2 \end{pmatrix};$$

the corresponding parameter values λ_1, λ_2 are, according to (3.43), subjected to the condition

$$\lambda_1(1+z_1\bar{z}_1)+\lambda_2(1+z_2\bar{z}_2) = 0.$$

Thus we may choose $\lambda_1 = -(1+z_2\bar{z}_2)$, $\lambda_2 = 1+z_1\bar{z}_1$. Then

(12.5) $$A = z_1\bar{z}_1 - z_2\bar{z}_2, \qquad B = (1+z_2\bar{z}_2)\bar{z}_1-(1+z_1\bar{z}_1)\bar{z}_2.$$

Now we have to transform this circle $\mathfrak{C}$ onto the unit circle. In view of (6.43) we determine a rotational Moebius transformation

$$\mathfrak{S} = \begin{pmatrix} a & b \\ -\bar{b} & \bar{a} \end{pmatrix}$$

such that

$$\mathfrak{S}'\mathfrak{C}\bar{\mathfrak{S}} = \lambda\begin{pmatrix} 1 & 0 \\ 0 & -1 \end{pmatrix} \qquad \text{(real } \lambda\text{)}.$$

This leads to the condition $2a\bar{b}A+a^2B-\bar{b}^2\bar{B} = 0$ or

$$\frac{A\bar{b}+Ba}{\bar{b}} = \frac{\bar{B}\bar{b}-Aa}{a} = \rho,$$

which is thus seen to be the eigenvalue problem of the hermitian matrix $\mathfrak{C}$:

$$A\bar{b}+Ba = \rho\bar{b}$$

$$\bar{B}\bar{b}-Aa = \rho a.$$

The characteristic values (roots) of $\mathfrak{C}$ are

(12.51) $$\rho = \pm\sqrt{(A^2+B\bar{B})} = \pm Ar$$

where r is the radius of the circle $\mathfrak{C}$. Thus we may take

(12.52) $$a = A(1+r) = A-\sqrt{(A^2+B\bar{B})}, \qquad b = -\bar{B},$$

and

(12.53) $$\mathfrak{R} = \mathfrak{S}^{-1} = \begin{pmatrix} A-\sqrt{(A^2+B\bar{B})} & \bar{B} \\ -B & A+\sqrt{(A^2+B\bar{B})} \end{pmatrix}.$$

3. The spherical transformation corresponding to a given rotational Moebius transformation

$$\mathfrak{R} = \begin{pmatrix} u & v \\ -\bar{v} & \bar{u} \end{pmatrix}, \qquad |\mathfrak{R}| = u\bar{u}+v\bar{v} = 1, \tag{12.6}$$

is a rotation of the sphere about **O**. It can be represented by a proper orthogonal matrix R whose elements depend on the two complex numbers

$$u = u_1+iu_2, \qquad v = v_1+iv_2.$$

The transformation defined by R maps great circles Γ onto great circles Γ^* while the Moebius transformation $\mathfrak{R}$ acts on the corresponding circles in the z-plane, changing the circle $\mathfrak{C} = \begin{pmatrix} A & B \\ \bar{B} & -A \end{pmatrix}$ onto $\mathfrak{C}^* = \begin{pmatrix} A^* & B^* \\ \bar{B}^* & -A^* \end{pmatrix}$. Putting $\mathfrak{S} = \mathfrak{R}^{-1} = \begin{pmatrix} \bar{u} & -v \\ \bar{v} & u \end{pmatrix}$, we have according to (6.42), $\mathfrak{C}^* = \mathfrak{S}'\mathfrak{C}\mathfrak{S}$, which in explicit form means that

$$\begin{cases} A^* = (u\bar{u}-v\bar{v})A+\bar{u}vB+u\bar{v}\bar{B} \\ B^* = -2\bar{u}\bar{v}A+\bar{u}^2B-\bar{v}^2\bar{B} \\ \overline{B^*} = -2uvA-v^2B+u^2\bar{B}. \end{cases} \tag{12.61}$$

Now let a, b, c and a^*, b^*, c^* be the coordinates of the planes of the two stereographically corresponding great circles Γ, Γ^* respectively. Then by (3.41) and (3.42)

$$\begin{pmatrix} a^* \\ b^* \\ c^* \end{pmatrix} = \begin{pmatrix} 0 & 1 & 1 \\ 0 & i & -i \\ -2 & 0 & 0 \end{pmatrix} \begin{pmatrix} A^* \\ B^* \\ \overline{B^*} \end{pmatrix} = \begin{pmatrix} 0 & 1 & 1 \\ 0 & i & -i \\ -2 & 0 & 0 \end{pmatrix} \begin{pmatrix} u\bar{u}-v\bar{v} & \bar{u}v & u\bar{v} \\ -2\bar{u}\bar{v} & \bar{u}^2 & -\bar{v}^2 \\ -2uv & -v^2 & u^2 \end{pmatrix} \begin{pmatrix} A \\ B \\ \bar{B} \end{pmatrix}$$

$$= \begin{pmatrix} 0 & 1 & 1 \\ 0 & i & -i \\ -2 & 0 & 0 \end{pmatrix} \begin{pmatrix} u\bar{u}-v\bar{v} & \bar{u}v & u\bar{v} \\ -2\bar{u}\bar{v} & \bar{u}^2 & -\bar{v}^2 \\ -2uv & -v^2 & u^2 \end{pmatrix} \begin{pmatrix} 0 & 0 & -\frac{1}{2} \\ \frac{1}{2} & -\frac{1}{2}i & 0 \\ \frac{1}{2} & \frac{1}{2}i & 0 \end{pmatrix} \begin{pmatrix} a \\ b \\ c \end{pmatrix} = R \begin{pmatrix} a \\ b \\ c \end{pmatrix}$$

where

$$R = \begin{pmatrix} \frac{1}{2}(u^2+\bar{u}^2-v^2-\bar{v}^2) & \frac{1}{2}i(u^2-\bar{u}^2+v^2-\bar{v}^2) & \bar{u}\bar{v}+uv \\ \frac{1}{2}i(\bar{u}^2-u^2+v^2-\bar{v}^2) & \frac{1}{2}(u^2+\bar{u}^2+v^2+\bar{v}^2) & i(\bar{u}\bar{v}-uv) \\ -(\bar{u}v+u\bar{v}) & i(\bar{u}v-u\bar{v}) & u\bar{u}-v\bar{v} \end{pmatrix}$$

$$= \begin{pmatrix} u_1^2-u_2^2-v_1^2+v_2^2 & -2(u_1u_2+v_1v_2) & 2(u_1v_1-u_2v_2) \\ +2(u_1u_2-v_1v_2) & u_1^2-u_2^2+v_1^2-v_2^2 & 2(u_1v_2+u_2v_1) \\ -2(u_1v_1+u_2v_2) & -2(u_1v_2-u_2v_1) & u_1^2+u_2^2-v_1^2-v_2^2 \end{pmatrix}. \tag{12.62}$$

In virtue of the normalization (12.6), viz. $u\bar{u}+v\bar{v} = 1$, this is the well-known representation of the rotation group by means of the so-called 'quaternion parameters'[2] $u_1, -v_2, v_1, u_2$.

For the geometrical characterization of the rotation represented by the matrix (12.62) we have to determine its axis and its angle. It is readily established that the axis coincides with the diameter through the point with the coordinates $-v_2, v_1, u_2$; indeed

$$R\begin{pmatrix}-v_2\\ v_1\\ u_2\end{pmatrix} = \begin{pmatrix}-v_2\\ v_1\\ u_2\end{pmatrix}.$$

The rotation angle ψ is the one shown in the normal form of the initial Moebius transformation $\mathfrak{R}$ [cf. (12.23)]. This normal form is obtained in the form $\mathfrak{T}\mathfrak{R}\mathfrak{T}^{-1}$ where the matrix $\mathfrak{T} = \begin{pmatrix}1 & -\gamma\\ \bar{\gamma} & 1\end{pmatrix}$ represents the Moebius transformation which maps one of the fixed points, γ, of the Moebius transformation $\mathfrak{R}$ onto 0, so that the corresponding rotation R is transformed into a rotation about the polar axis **NS** of the sphere. Thus γ is a root of the quadratic equation $\bar{v}\gamma^2+(u-\bar{u})\gamma+v = 0$, and we obtain

$$\psi = 2\,\mathrm{arc}(\bar{u}\gamma\bar{\gamma}+\bar{v}\gamma-v\bar{\gamma}+u). \tag{12.63}$$

Remark. The rotational Moebius transformations leave the imaginary unit circle $\begin{pmatrix}1 & 0\\ 0 & 1\end{pmatrix}$ invariant and they are entirely characterized by this property. Correspondingly the rotations in space leave invariant the plane with the coordinates

$$a = b = c = 0, \qquad d = 2$$

that is, the plane at infinity ($x_4 = 0$) in the homogeneous coordinates x_1, x_2, x_3, x_4 where

$$\xi = \frac{x_1}{x_4}, \qquad \eta = \frac{x_2}{x_4}, \qquad \zeta = \frac{x_3}{x_4}.$$

But this property is not sufficient to characterize the space rotations. They are those projective transformations which leave the unit sphere and the infinite plane invariant.

4. In a similar manner we may explicitly determine the projective transformation S in space which corresponds to a Moebius transformation $\mathfrak{H}$ leaving invariant the real unit circle (cf. § 12, **a**). Any such transformation S leaves the unit sphere and the equatorial plane invariant, mapping the upper half of the sphere onto itself. Every spherical circle orthogonal to the equator is mapped onto a

[2] Cf. L.A., p. 239, equation (28, 29)". Cf. also Cayley [1] and Watson [1].

circle of the same kind; therefore every plane parallel to the polar axis is mapped onto a plane of the same kind whose distance from the axis remains less than, equal to, or greater than unity.

Now consider the equatorial plane as a projective plane. In its finite parts it coincides with the z-plane. The space transformation S is uniquely defined by a plane projective transformation which leaves invariant the unit circle $\xi^2+\eta^2 = 1$, mapping the interior of this circle onto itself.

The given Moebius transformation $\mathfrak{H} = \begin{pmatrix} u & v \\ \bar v & \bar u \end{pmatrix}$ [cf. (12.1)] may be assumed to be normalized so that

$$|\mathfrak{H}| = u\bar u - v\bar v = 1. \tag{12.7}$$

It will map the circle $\mathfrak{C} = \begin{pmatrix} A & B \\ \bar B & A \end{pmatrix}$ (which is orthogonal to the unit circle $\begin{pmatrix} 1 & 0 \\ 0 & -1 \end{pmatrix}$, cf. § 4, ex. 7) onto the circle

$$\mathfrak{C}^* = \mathfrak{S}'\mathfrak{C}\bar{\mathfrak{S}} = \begin{pmatrix} A^* & B^* \\ \overline{B^*} & A^* \end{pmatrix} \qquad \left[\mathfrak{S} = \mathfrak{H}^{-1} = \begin{pmatrix} \bar u & -v \\ -\bar v & u \end{pmatrix}\right]$$

whose coordinates are given by

$$\begin{cases} A^* = (u\bar u+v\bar v)A-\bar u vB-u\bar v\bar B \\ B^* = -2\bar u\bar vA+\bar u^2B+\bar v^2\bar B \\ C^* = -2uvA+v^2B+u^2\bar B. \end{cases} \tag{12.71}$$

Proceeding to plane coordinates a, b, c, d and a^*, b^*, c^*, d^*, by (3.41) and (3.42) we notice that $c = 0$, $c^* = 0$ (for the planes in question are perpendicular to the equatorial plane) and

$$\begin{pmatrix} a^* \\ b^* \\ d^* \end{pmatrix} = \begin{pmatrix} 0 & 1 & 1 \\ 0 & i & -i \\ 2 & 0 & 0 \end{pmatrix} \begin{pmatrix} u\bar u+v\bar v & -\bar uv & -u\bar v \\ -2\bar u\bar v & \bar u^2 & \bar v^2 \\ -2uv & v^2 & u^2 \end{pmatrix} \begin{pmatrix} 0 & 0 & \frac12 \\ \frac12 & \frac12 i & 0 \\ \frac12 & \frac12 i & 0 \end{pmatrix} \begin{pmatrix} a \\ b \\ d \end{pmatrix} = \tilde S \begin{pmatrix} a \\ b \\ d \end{pmatrix}$$

where

$$\tilde S = \begin{pmatrix} \frac12(u^2+\bar u^2+v^2+\bar v^2) & \frac12 i(u^2-\bar u^2+\bar v^2-v^2) & uv+\bar u\bar v \\ \frac12 i(\bar u^2-u^2-v^2+\bar v^2) & \frac12(u^2+\bar u^2-v^2-\bar v^2) & i(\bar u\bar v-uv \\ -(\bar uv+u\bar v) & i(\bar uv-u\bar v) & u\bar u+v\bar v \end{pmatrix}$$

$$= \begin{pmatrix} u_1^2-u_2^2+v_1^2-v_2^2 & -2(u_1u_2-v_1v_2) & 2(u_1v_1-u_2v_2) \\ 2(u_1u_2+v_1v_2) & u_1^2-u_2^2-v_1^2+v_2^2 & 2(u_1v_2+u_2v_1) \\ -2(u_1v_1+u_2v_2) & -2(u_1v_2-u_2v_1) & u_1^2+u_2^2+v_1^2+v_2^2 \end{pmatrix}. \tag{12.72}$$

Now we introduce the homogeneous coordinates x_1, x_2, x_3, x_4 as in (12.64). Then

$$ax_1+bx_2+dx_4 = 0 \tag{12.73}$$

is the equation of a line in the $\xi\eta$-plane (as well as the equation of a plane in space parallel to the ζ-axis). The matrix, as representative of a projective transformation in line coordinates a, b, d, is a three-rowed Lorentz matrix, that is,

$$\tilde{S}'\tilde{L}\tilde{S} = \tilde{L}, \qquad \tilde{L} = \begin{pmatrix} 1 & 0 & 0 \\ 0 & 1 & 0 \\ 0 & 0 & -1 \end{pmatrix}. \tag{12.74}$$

Accordingly the corresponding projective transformation in point coordinates x_1, x_2, x_4, viz.

$$\begin{pmatrix} y_1 \\ y_2 \\ y_4 \end{pmatrix} = \tilde{S}_1 \begin{pmatrix} x_1 \\ x_2 \\ x_4 \end{pmatrix} \tag{12.75}$$

will have as its matrix $\tilde{S}_1$ the contragredient[3] matrix $\tilde{S}'^{-1}$ of $\tilde{S}$ which with regard to (12.74) is obtained by replacing in $\tilde{S}$ the elements of the third row and of the third column, excepting their common element, by the numerically opposite values.[4] (Cf. § 14, ex. 2.)

The projective space transformation corresponding to the Moebius transformation $\mathfrak{H}$ in the z-plane, and to the projective transformation $\tilde{S}_1$ in the $\xi\eta$-plane, will be found in the following way. The coordinates x_1, x_2, x_3, x_4 of a point x on the sphere satisfy the condition $x_3^2 = x_4^2-x_1^2-x_2^2$, and taking into account (12.7) we have by (12.75)

$$y_3^2 = y_4^2-y_1^2-y_2^2 = x_3^2.$$

Hence $y_3 = \pm x_3$; but since the upper half of the sphere is mapped onto the upper half, $y_3 = x_3$. Hence the transformation which maps the point x onto the point y of the sphere is given by

$$y = Sx$$

where

$$S = \begin{pmatrix} u_1^2-u_2^2+v_1^2-v_2^2 & -2(u_1u_2-v_1v_2) & 0 & -2(u_1v_1-u_2v_2) \\ 2(u_1u_2+v_1v_2) & u_1^2-u_2^2-v_1^2+v_2^2 & 0 & -2(u_1v_2+u_2v_1) \\ 0 & 0 & 1 & 0 \\ 2(u_1v_1+u_2v_2) & 2(u_1v_2-u_2v_1) & 0 & u_1^2+u_2^2+v_1^2+v_2^2 \end{pmatrix}. \tag{12.76}$$

This is a parametric representation of the plane projective transformations which leave the unit circle and its interior invariant.

[3] Cf. L.A., p. 264, in particular equation (A, 8)′ and L.A., p. 198.

[4] This also becomes evident if we observe that the pole of the line (12.73) with respect to the circle $x_1^2+x_2^2-x_4^2 = 0$ has the coordinates $x_1 = a$, $x_2 = b$, $x_4 = -d$.

5. By the same method a parametric representation of those projective transformations which map the sphere onto itself and which correspond to plane euclidean motions can be found. In this case the interchanged circles on the sphere are those corresponding to the straight lines in the z-plane. Their hermitian matrices are of the form $\mathfrak{C} = \begin{pmatrix} 0 & B \\ \bar{B} & D \end{pmatrix}$ $(A = 0)$. Thus by (3.42) $d = c$. If $\mathfrak{H} = \begin{pmatrix} e^{i\alpha} & v \\ 0 & 1 \end{pmatrix}$ represents the given motion let

$$e^{i(\alpha/2)}\mathfrak{H}^{-1} = \begin{pmatrix} e^{-i(\alpha/2)} & -e^{-i(\alpha/2)}e^{-i(\alpha/2)}v \\ 0 & e^{i(\alpha/2)} \end{pmatrix} = \mathfrak{S};$$

we find the elements of the matrix $\mathfrak{C}^* = \mathfrak{S}'\mathfrak{C}\bar{\mathfrak{S}}$ to be

$$B^* = e^{-i\alpha}B, \qquad \overline{B^*} = e^{i\alpha}\bar{B}, \qquad D^* = -ve^{-i\alpha}B - \bar{v}e^{i\alpha}\bar{B} + D.$$

If $v = re^{i\beta}$ $(|v| = r)$, then

$$\begin{pmatrix} a^* \\ b^* \\ c^* \end{pmatrix} = \begin{pmatrix} \cos\alpha & -\sin\alpha & 0 \\ \sin\alpha & \cos\alpha & 0 \\ -r\cos(\beta-\alpha) & -r\sin(\beta-\alpha) & 1 \end{pmatrix} \begin{pmatrix} a \\ b \\ c \end{pmatrix}.$$

This matrix can now be completed, in one and only one way, into a Lorentz matrix

$$T = \begin{pmatrix} \cos\alpha & -\sin\alpha & r\cos\beta & -r\cos\beta \\ \sin\alpha & \cos\alpha & r\sin\beta & -r\sin\beta \\ -r\cos(\beta-\alpha) & -r\sin(\beta-\alpha) & 1-\frac{1}{2}r^2 & \frac{1}{2}r^2 \\ -r\cos(\beta-\alpha) & -r\sin(\beta-\alpha) & -\frac{1}{2}r^2 & 1+\frac{1}{2}r^2 \end{pmatrix}$$

which represents the transformation in plane coordinates a, b, c, d. The matrix S representing the same transformation in point coordinates x_1, x_2, x_3, x_4 will then be the contragredient matrix T'^{-1} (cf. ex. 4), that is,

$$S = LTL,$$

which is obtained from T by replacing the elements of the fourth row and the fourth column, excepting the common element $(1+\frac{1}{2}r^2)$, by the numerically opposite values.

6. A two-point invariant (distance function) for the group $\mathscr{R}$ is given by the cross ratio

$$d_1(z_1, z_2) = \left(z_1, z_2; -\frac{1}{\bar{z}_1}, -\frac{1}{\bar{z}_2}\right) \tag{12.8}$$

[cf. (12.42)]

7. The group $\mathscr{E}$ of the euclidean displacements (12.3) is an invariant subgroup of the group $\mathscr{I}$ of all integral Moebius transformations, and the quotient group $\mathscr{I}/\mathscr{E}$ is isomorphic to the group of all proper dilatations $Z = az$, $(a > 0)$.

8. The group $\mathscr{G}^{(\rho^2)}$ of all Moebius transformations mapping the circle $(0, \rho)$ onto itself is represented by the matrices

$$\mathfrak{H} = \begin{pmatrix} a & b \\ \dfrac{\bar{b}}{\rho^2} & \bar{a} \end{pmatrix} \tag{12.9}$$

where ρ is real or pure imaginary. Then $\mathscr{G}^{(1)} = \mathscr{U}$, $\mathscr{G}^{\infty} = \mathscr{E}$, $\mathscr{G}^{(-1)} = \mathscr{R}$. Thus the groups $\mathscr{G}^{(\rho^2)}$ are similar to $\mathscr{U}$ if $\rho^2 > 0$, similar to $\mathscr{R}$ if $\rho^2 < 0$.

It may be observed that for pure imaginary b the matrix (12.9) is a normalized circle matrix (cf. § 9, ex. 1).

§ 13. The geometry of a transformation group

a. *Euclidean geometry.* As a preparation to the following exposition of the non-euclidean geometries it will first be shown how the essentials of euclidean geometry can be derived from certain properties of the group $\mathscr{E}$ of all euclidean displacements (motions).

Let us denote plane geometrical figures such as segments, triangles, lines, circles, etc., by $\mathscr{F}$, $\mathscr{F}_1$, $\mathscr{F}_2$. Euclidean geometry in the plane is the study of those properties of such figures $\mathscr{F}$ which remain unchanged if $\mathscr{F}$ is displaced within the plane in an arbitrary way. Distances between different points of a figure $\mathscr{F}$ and angles between lines or circles are not changed by displacement; the distance of a point of $\mathscr{F}$ from a fixed point in the plane does not remain unchanged.

If a figure $\mathscr{F}_1$ can be brought into coincidence with $\mathscr{F}_2$ by a displacement, then $\mathscr{F}_1$ and $\mathscr{F}_2$ are said to be 'congruent with respect to the group of all displacements' or briefly congruent. Thus the group $\mathscr{E}$ of all displacements (12.3) determines congruence of geometrical figures. Certain operations, elements of the group $\mathscr{E}$, enable us to test the congruence of two given figures.

Moreover, by its one-parametric subgroups, that is by certain group-structural properties, the group $\mathscr{E}$ defines the fundamental elements of euclidean geometry, the straight lines and the circles. They correspond to the two different types of one-parametric subgroups in $\mathscr{E}$.

1. *Parabolic subgroups*, obtained by continuous iteration of parabolic transformations of the group $\mathscr{E}$. All parabolic transformations of the group are translations $\mathfrak{T}_b$:

$$Z = \mathfrak{T}_b(z) = z + b.$$

For constant z and real variable s, the point

$$Z = \mathfrak{T}_b{}^s(z) = z + sb$$

travels on the well-defined straight line through z, parallel to the translation vector b (cf. § 10, ex. 9, and § 10, **f**, Theorem E).

2. *Elliptic subgroups.* Any non-translation in the displacement group $\mathscr{E}$ is elliptic. Let $\alpha \neq 2n\pi$; then

$$\gamma = \frac{b}{1-e^{i\alpha}}$$

is the only finite fixed point of the rotation (12.3), which can be written in the form

$$Z = \mathfrak{H}(z) = \gamma + e^{i\alpha}(z-\gamma).$$

Thus, given a constant $z \neq \gamma$, the point

$$Z = \mathfrak{H}^s(z) = \gamma + e^{i\alpha s}(z-\gamma)$$

will travel on the *circle* through z about γ if s is a real variable.

Every point z can be shifted into the position of any other point in the finite plane by a suitable displacement $\mathfrak{H}$. Hence if a function $f(z)$ is an invariant, that is, if

(13.1) $$f[\mathfrak{H}(z)] = f(z) \qquad \text{for all } \mathfrak{H} \text{ in } \mathscr{E},$$

then it is necessarily constant. This is what we mean when we say that there is no one-point invariant of the displacement group. This fact is of course a consequence of the transitivity of the group $\mathscr{E}$.

Next to the point z we consider the segment, defined by a pair of points z_1, z_2. There is a function of the two points z_1, z_2 which is invariant under any displacement; this is their distance $|z_1-z_2|$. Naturally, any function of the distance is invariant. Apart from this arbitrariness in the choice of a function it will be seen, however, that *the distance* $|z_1-z_2|$ *is the only displacement invariant attached to two points.* This means that every invariant function $f(z_1, z_2)$ is a function of the distance $|z_1-z_2|$.

Indeed, if $f(z_1, z_2)$ is an invariant it will not change its value when z_1, z_2 are both subjected to an arbitrary translation: $f(z_1+b, z_2+b) = f(z_1, z_2)$. For $b = -z_2$ it follows

$$f(z_1, z_2) = f(z_1-z_2, 0).$$

Thus the invariant is a function of the difference z_1-z_2.

But $f(z_1, z_2)$ retains its value if z_1, z_2 are subjected to an arbitrary rotation:

$$f(z_1, z_2) = f[e^{i\alpha}(z_1-z_2), 0].$$

Let $\alpha = -\operatorname{arc}(z_1-z_2)$; then

$$f(z_1, z_2) = f(|z_1-z_2|, 0) \qquad \text{q.e.d.}$$

Remark. The simple, not two-fold transitivity of the group $\mathscr{E}$ is essential for the existence and uniqueness of the invariant distance (cf. § 13, **c**). In fact it is easy to see that a two-fold transitive group of transformations cannot have an invariant attached to two points, and that an intransitive group may have several invariants of two points.

b. *$\mathcal{G}$-geometry.* It has been the basic idea of Klein's 'Erlanger Programm' (1872) to associate a geometry—we shall say a $\mathcal{G}$-geometry—with a rather arbitrary group $\mathcal{G}$ of transformations operating in the 'space' of this geometry the way the group $\mathcal{E}$ of euclidean motions operates in the euclidean plane. The 'space' of the $\mathcal{G}$-geometry may be defined as a domain of simple transitivity of the given group $\mathcal{G}$ of transformations.

We shall study in particular those plane geometries which in Klein's sense belong to the groups of the hyperbolic, parabolic, and the elliptic bundle (cf. § 12). These bundles may be assumed in their normal forms. Thus $\mathcal{G}$ will be one of the groups $\mathcal{U}_+$, $\mathcal{E}$, and $\mathcal{R}$ respectively, with the interior of the unit circle, the ordinary euclidean plane, and the completed plane of the complex numbers as domain of simple transitivity. In each case this domain may be denoted by $\mathcal{D}$; it is the plane (or 'space') of the corresponding $\mathcal{G}$-geometry.

As in the case of the euclidean geometry (§ 13, **a**) two figures in $\mathcal{D}$ will be said to be $\mathcal{G}$-congruent if the one can be mapped onto the other by a transformation of $\mathcal{G}$ (cf. ex. 6).

The points of a $\mathcal{G}$-geometry are the complex numbers within $\mathcal{D}$. The other fundamental elements of the $\mathcal{G}$-geometry, called straight lines, cycles, and circles, are introduced by means of the one-parametric subgroups of $\mathcal{G}$ which consist of the continuous iterations $\mathfrak{H}^s$ of an element $\mathfrak{H}$ of $\mathcal{G}$, as constructed in § 10, **f**. From § 10, **f**, Theorem E it is known that every curve $\mathfrak{C}$ which is obtained by applying all the powers of $\mathfrak{H}$ to a certain point z_0 of $\mathcal{D}$, is a circle or an arc of a circle.[5] Its parametric representation is given by the equation

$$z = \mathfrak{H}^s(z_0) \qquad (-\infty < s < \infty). \tag{13.1}$$

Definitions. The circle $\mathfrak{C}$ is said to be a $\mathcal{G}$-circle if the transformation $\mathfrak{H}$ is elliptic. A fixed point γ of $\mathfrak{H}$, situated within $\mathcal{D}$, is called a centre of the $\mathcal{G}$-circle. The circle arc (13.1) is called a $\mathcal{G}$-hypercycle if $\mathfrak{H}$ is proper hyperbolic; it is called a $\mathcal{G}$-horocycle if $\mathfrak{H}$ is parabolic. (Cf. § 10, **f**, Theorem E and Corollary 1).

If $\mathcal{G} = \mathcal{R}$, every transformation in $\mathcal{G}$ is elliptic and therefore there are no $\mathcal{G}$-cycles. If $\mathcal{G} = \mathcal{E}$, every transformation is elliptic or parabolic and there are no hypercycles.

$\mathcal{G}$-circles and $\mathcal{G}$-cycles which lie on ordinary (euclidean) straight lines through 0 are, by definition, '$\mathcal{G}$-straight lines' through 0. Every curve congruent to a $\mathcal{G}$-straight line through 0 is called a $\mathcal{G}$-straight line.

Thus a $\mathcal{U}_+$-straight line is a $\mathcal{U}_+$-hypercycle orthogonal to (and bounded by) the unit circle. An $\mathcal{E}$-straight line is a horocycle; this is a circle through the point at infinity, that is, an ordinary straight line.

It will be shown how, on the basis of the preceding definitions, geometrical theories can be developed. The case $\mathcal{G} = \mathcal{E}$ has been indicated in § 13, **a**; it

[5] It is the path curve of z_0 under the transformation group consisting of the powers $\mathfrak{H}^s$ of $\mathfrak{H}$.

will not be further developed. For $\mathscr{G} = \mathscr{U}_+$ we obtain 'hyperbolic geometry,'[6] and for $\mathscr{G} = \mathscr{R}$: 'spherical geometry.' In each of these geometries the elements of the group $\mathscr{G}$ play the role of motions in euclidean geometry; therefore we call $\mathscr{U}_+$ the group of the hyperbolic motions and $\mathscr{R}$ the group of the spherical motions.

c. *Distance function.* It has been shown (§ 12, **e** and ex. 6) how a distance function can be constructed for hyperbolic and spherical geometry. We shall now develop a second method common to the three groups $\mathscr{U}_+$, $\mathscr{E}$, $\mathscr{R}$, and possibly other groups $\mathscr{G}$. At the same time the uniqueness of the distance function will be established.

The proof will be based on a few simple properties of the group $\mathscr{G}$ which are realized in the cases $\mathscr{U}_+$, $\mathscr{E}$, $\mathscr{R}$.

I. The elements of $\mathscr{G}$ are mappings of $\mathscr{D}$ onto itself.

As group elements they have their inverses in $\mathscr{G}$ and are therefore one-to-one and invertible.

II. $\mathscr{G}$ is simply, not two-fold, transitive.

All the elements $\mathfrak{H}_0$ of $\mathscr{G}$ for which a point z_0 of $\mathscr{D}$ is a fixed point, form a subgroup $\mathscr{H}_0$ of $\mathscr{G}$ which is called the 'stability subgroup' of $\mathscr{G}$ at z_0.

Theorem A. *The stability subgroups at the points in $\mathscr{D}$ form a complete set of conjugate subgroups of $\mathscr{G}$.*

Proof. Let z_1 be a point in $\mathscr{D}$ and $\mathscr{H}_1$ the stability subgroup at z_1. With regard to the assumption II there is a transformation $\mathfrak{T}$ in $\mathscr{G}$ such that $z_1 = \mathfrak{T}(z_0)$. Then

$$\mathfrak{T}\mathfrak{H}_0\mathfrak{T}^{-1}(z_1) = \mathfrak{T}\mathfrak{H}_0(z_0) = \mathfrak{T}(z_0) = z_1,$$

which means that for every $\mathfrak{H}_0$ in $\mathscr{H}_0$ the transformation $\mathfrak{T}\mathfrak{H}_0\mathfrak{T}^{-1}$ is an element of $\mathscr{H}_1$. Similarly for every $\mathfrak{H}_1$, in $\mathscr{H}_1$ the transformation $\mathfrak{T}^{-1}\mathfrak{H}_1\mathfrak{T}$ is an element of $\mathscr{H}_0$. In the symbolism of group theory we have therefore $\mathscr{H}_1 = \mathfrak{T}\mathscr{H}_0\mathfrak{T}^{-1}$, which means that $\mathscr{H}_0$ and $\mathscr{H}_1$ are conjugate subgroups.

From § 12, **a** and **b**, and § 13, **a**, it is evident that in each of the groups $\mathscr{U}_+$, $\mathscr{E}$, $\mathscr{R}$ the stability subgroup at the point 0 is the group of all rotations about this point. Therefore we assume that $\mathscr{D}$ contains the point 0 and

III. The stability subgroup of $\mathscr{G}$ at 0 is the group of all (euclidean) rotations about 0.

On the basis of these three assumptions we shall now construct an invariant $f(z_1, z_2)$ defined for all pairs of points z_1, z_2 in $\mathscr{D}$. Let $\mathfrak{H}$ be an element of $\mathscr{G}$ and

$$\mathfrak{H}(z_1) = Z_1, \qquad \mathfrak{H}(z_2) = Z_2.$$

[6] It is sometimes called 'non-euclidean geometry,' a name that we shall use in a wider sense to include the geometries to be dealt with in § 15. It is also called Lobačevski geometry.

Choose an arbitrary point z_0 in $\mathscr{D}$ and find two transformations $\mathfrak{H}_{z_2}$, $\mathfrak{H}_{Z_2}$ in $\mathscr{G}$ such that

$$\mathfrak{H}_{z_2}(z_2) = \mathfrak{H}_{Z_2}(Z_2) = z_0.$$

Then z_0 is a fixed point of the transformation

$$\mathfrak{H}_0 = \mathfrak{H}_{Z_2}\mathfrak{H}\mathfrak{H}_{z_2}^{-1}.$$

If we apply this transformation to $z = \mathfrak{H}_{z_2}(z_1)$, we get

$$\mathfrak{H}_0[\mathfrak{H}_{z_2}(z_1)] = \mathfrak{H}_{Z_2}[\mathfrak{H}(z_1)] = \mathfrak{H}_{Z_2}(Z_1). \tag{13.2}$$

If the stability subgroup $\mathscr{H}_0$ at z_0 consists of the unit element only, so that necessarily $\mathfrak{H}_0 = \mathfrak{E}$, then by (13.2) $\mathfrak{H}_{z_2}(z_1)$ is an invariant of the group $\mathscr{G}$.

But now let us take $z_0 = 0$; according to III the stability subgroup $\mathscr{H}_0$ is then the group of all rotations about 0. Thus

$$\mathfrak{H}_0(z) = e^{i\alpha}z,$$

and from (13.2) it is evident that

$$f(z_1, z_2) = |\mathfrak{H}_{z_2}(z_1)| \tag{13.21}$$

represents a two-point invariant of the group $\mathscr{G}$.

We observe that for all z in $\mathscr{D}$

$$f(z, 0) = |\mathfrak{H}_0(z)| = |z| \tag{13.22}$$

is the only invariant of the stability group $\mathscr{H}_0$. Any other invariant of 0 and z is therefore a function of $|z|$. Thus as long as the symmetry of the function $f(z_1, z_2)$ has not been established, we can only say that there is a function $F(r)$ of the non-negative variable r such that

$$f(0, z) = F(|z|).$$

Moreover if for an element $\mathfrak{H}$ of $\mathscr{G}$ we have

$$\mathfrak{H}(0) = Z_0, \qquad \mathfrak{H}(z) = Z,$$

then

$$f(Z, Z_0) = |z| \tag{13.23}$$

is uniquely defined; similarly $f(Z_0, Z) = F(|z|)$.

Now let $f_1(z_1, z_2)$ be another two-point invariant of $\mathscr{G}$. The invariant of the subgroup $\mathscr{H}_0$ being unique, it follows that there is a function $F_1(r)$ such that $f_1(z, 0) = F_1(|z|)$; hence

$$f_1(Z, Z_0) = F_1(|z|) = F_1[f(Z, Z_0)],$$

where Z_0 is defined by the choice of $\mathfrak{H}$. Since z is an arbitrary point in $\mathscr{D}$, the same is true for Z; thus the last equation is valid for all pairs of points Z_0, Z in $\mathscr{D}$. Therefore

Theorem B. *For a group $\mathscr{G}$ that satisfies the conditions I–III there is one and only one independent two-point invariant $f(z_1, z_2)$. It is given by* (13.21) *and may be called the $\mathscr{G}$-distance function of z_1, z_2.*

It is readily verified that the $\mathscr{E}$-distance function of any two points z_1, z_2 in the euclidean plane is equal to the euclidean distance $|z - z_2|$. Some of the obvious properties of this distance are also properties of the $\mathscr{G}$-distance function. The function $f(z_1, z_2)$ given by (13.21) is non-negative and equal to zero if and only if $z_2 = z_1$. Indeed $\mathfrak{H}_{z_2}(z_2) = z_0 = 0$, and because the transformation $\mathfrak{H}_{z_2}$ possesses a unique inverse in $\mathscr{G}$ it follows that the equation $\mathfrak{H}_{z_2}(z) = 0$ has the unique solution $z = z_2$.

Without further assumptions concerning the group $\mathscr{G}$ it seems impossible to prove that the distance function $f(z_1, z_2)$ is symmetric, that is $f(z_2, z_1) = f(z_1, z_2)$. But symmetry follows if we assume that

$$f(0, z) = f(z, 0) \tag{13.3}$$

which for the group $\mathscr{G}$ involves the following condition:

IV. A transformation $\mathfrak{H}_{z_1}$ in $\mathscr{G}$ for which $\mathfrak{H}_{z_1}(z_1) = 0$ has the property $|\mathfrak{H}_{z_1}(0)| = |z_1|$ [cf. (13.23)].

Indeed, making use of the invariance of $f(z_1, z_2)$, we can prove the symmetry of this function.

$$\begin{aligned} f(z_1, z_2) &= |\mathfrak{H}_{z_2}(z_1)| \\ &= f[\mathfrak{H}_{z_1}(z_1), \mathfrak{H}_{z_1}(z_2)] = f[0, \mathfrak{H}_{z_1}(z_2)] \\ &= f[\mathfrak{H}_{z_1}(z_2), 0] = |\mathfrak{H}_{z_1}(z_2)| \\ &= f(z_2, z_1). \end{aligned}$$

Now let $\mathscr{G} = \mathscr{U}_+$, or $\mathscr{E}$, or $\mathscr{R}$ and $\varepsilon = -1$, or 0, or $+1$ respectively. The elements of $\mathscr{G}$ are then the transformations $\mathfrak{H}(z) = (az+b)/(-\varepsilon\bar{b}z+\bar{a})$, and $\mathfrak{H}_{z_2}(z_2) = 0$ implies $b = -az_2$. Thus by (13.21) we have the distance functions

$$f_\varepsilon(z_1, z_2) = \frac{|z_1 - z_2|}{|1+\varepsilon\bar{z}_2 z_1|}; \tag{13.31}$$

these are readily seen to be symmetric. They are naturally connected with the functions d_{-1}, d_1 which have been introduced in (12.42) and § 12, ex. 6; cf. also § 13, ex. 5.

d. *$\mathscr{G}$-circles.* Let $\mathscr{G}$ be again one of the three groups $\mathscr{U}_+$, $\mathscr{E}$, $\mathscr{R}$. As in elementary euclidean geometry a circle in $\mathscr{G}$-geometry can be characterized as follows:

Theorem C. *A $\mathscr{G}$-circle is the locus of all the points in $\mathscr{D}$ which have constant $\mathscr{G}$-distance from a point z_0 in $\mathscr{D}$ and every curve with this property is a $\mathscr{G}$-circle. The point z_0 is called a $\mathscr{G}$-centre of the $\mathscr{G}$-circle.*

Proof. By definition (§ 13, **b**) a $\mathscr{G}$-circle consists of all the points $\mathfrak{H}^s(z_1)$ where z_1 is constant in $\mathscr{D}$ and $\mathfrak{H}$ is an elliptic element of $\mathscr{G}$. Thus $\mathfrak{H}$ has at least one fixed point z_0 in $\mathscr{D}$. By means of the transformation $\mathfrak{H}_{z_0}$ of $\mathscr{G}$ (mapping z_0 onto 0), the $\mathscr{G}$-circle is mapped onto the set of all the points

$$z^* = \mathfrak{H}_0^s(z_1^*) \qquad \text{where } z_1^* = \mathfrak{H}_{z_0}(z_1),$$

hence

$$\mathfrak{H}_0 = \mathfrak{H}_{z_0}\mathfrak{H}\mathfrak{H}_{z_0}^{-1}.$$

The transformation $\mathfrak{H}_0$ has 0 as fixed point and is therefore a rotation; hence $|z^*| = |z_1^*|$ and

$$f(z, z_0) = f(z_1, z_0). \tag{13.4}$$

Conversely, for a definite choice of z_0, z_1 in $\mathscr{D}$, (13.4) is equivalent to

$$\mathfrak{H}_{z_0}(z) = e^{is}\mathfrak{H}_{z_0}(z_1) \qquad (s \text{ real}) \tag{13.41}$$

whence (by operating with $\mathfrak{H}_{z_0}^{-1}$ on both sides of this relation) we obtain a parametric representation of the curve given by (13.4). By (10.76) this can be written in the form $z = \mathfrak{H}^s(z_1)$, where $\mathfrak{H}$ is an elliptic Moebius transformation. Thus (13.4) represents a $\mathscr{G}$-circle through z_1 about z_0 as $\mathscr{G}$-centre.

From (13.22) it is seen that $\mathscr{G}$-circles about 0 are ordinary circles about 0.

In the case $\mathscr{G} = \mathscr{U}_+$, that is, in hyperbolic geometry, $\mathscr{D}$ is the interior of the unit circle. Its perimeter about 0 will be called the 'horizon' or the 'absolute' of the hyperbolic plane. Its points represent the infinity of the 'plane' $\mathscr{D}$. It will be denoted by Γ.

Every hyperbolic disk (that is, interior of a $\mathscr{U}_+$-circle) consists of inner points of $\mathscr{D}$ only and therefore has no point in common with the horizon. Indeed every point of a hyperbolic disk is obtained from one of its points by a hyperbolic motion, that is, a transformation of $\mathscr{D}$ into itself. Applied to the points z_1 of Γ the formula (13.41) would suggest that every point z_0 in $\mathscr{D}$ is a $\mathscr{G}$-centre of this circle Γ. This demonstrates the peculiar position of the circle Γ. We avoid this irregularity by assuming $\mathscr{D}$ to be the open circular disk without its circumference.

In the case of euclidean geometry $\mathscr{D}$ is the plane of all complex numbers and Γ is the point circle at ∞. In the case of elliptic geometry ($\mathscr{G} = \mathscr{R}$) the horizon Γ is empty.

Examples

1. Any two stability subgroups of a group $\mathscr{G}$ have only the identity in common. The stability subgroups of $\mathscr{R}$ at two antipodal points are the same.

2. The proof of the existence and uniqueness of the distance function (Theorem B) can be extended to slightly more general conditions. It appears to be sufficient to assume instead of $\mathscr{D}$ a topological image $\tilde{\mathscr{D}}$ of $\mathscr{D}$ in the complex plane; that is, we assume the existence of a topological transformation $\mathfrak{T}$,

not necessarily an element of $\mathscr{G}$, such that if $\mathscr{H}_0$ is the stability subgroup of $\mathscr{G}$ at z_0 then $\mathfrak{T}\mathscr{H}_0\mathfrak{T}^{-1}$, the stability subgroup of $\mathfrak{T}\mathscr{G}\mathfrak{T}^{-1}$ at $\mathfrak{T}(z_0) = 0$, is the group of all rotations in the plane about 0; then for every $\mathfrak{H}_0$ in $\mathscr{H}_0$,

$$\mathfrak{T}\mathfrak{H}_0\mathfrak{T}^{-1}(z) = e^{i\alpha}z \qquad (\alpha \text{ real}).$$

Thus by applying $\mathfrak{T}$ on both sides of (13.2) we get

$$\mathfrak{T}[\mathfrak{H}_{z_2}(Z_1)] = \mathfrak{T}\mathfrak{H}_0\mathfrak{T}^{-1}\mathfrak{T}\mathfrak{H}_{z_2}(z_1) = e^{i\alpha}\mathfrak{T}\mathfrak{H}_{z_2}(z_1). \tag{13.5}$$

Hence

$$f(z_1, z_2) = |\mathfrak{T}\mathfrak{H}_{z_2}(z_1)| \tag{13.51}$$

represents the two-point invariant (distance function) of the group $\mathscr{G}$ operating in $\mathscr{D}$. The generalization (13.51) of (13.21) will be useful in the following example.

3. Let $\mathscr{G}$ be the group whose elements are products of an even number of inversions within the parabolic bundle of all circles passing through the point 0 of the completed plane. Every element of this group $\mathscr{G}$ appears in the form

$$Z = \frac{z}{bz+e^{i\alpha}} \qquad (\alpha \text{ real}) \tag{13.6}$$

and the domain $\mathscr{D}$ of the $\mathscr{G}$-geometry is the completed plane dotted at 0 (so that 0 does not belong to $\mathscr{D}$). The transformation (13.6) is parabolic if and only if $e^{i\alpha} = 1$. For $b = 0$ it represents a rotation about 0 (which is not a point of $\mathscr{D}$) and a 'rotation about ∞' (which is a point of $\mathscr{D}$). Thus, using the notation of § 13, c, we choose $z_0 = \infty$ and

$$\mathfrak{H}_{z_2} = \begin{pmatrix} 1 & 0 \\ -\dfrac{1}{z_2} & 1 \end{pmatrix}.$$

By the transformation

$$z^* = \mathfrak{T}(z) = \frac{1}{z}$$

we carry ∞ into 0. Then

$$\mathfrak{T}\mathfrak{H}_{z_2}(z_1) = \frac{1}{z_1} - \frac{1}{z_2},$$

and

$$f(z_1, z_2) = \left|\frac{1}{z_1} - \frac{1}{z_2}\right| \tag{13.61}$$

is the $\mathscr{G}$-distance function.

All $\mathscr{G}$-straight lines are obtained as horocycles, orbits of parabolic groups; by (10.73) they are found to be the ordinary circles through 0. The $\mathscr{G}$-geometry is essentially identical with euclidean geometry.

4. The symmetry of the general distance function $f(z_1, z_2)$ could not be proved. But by Theorem B we know that there is a function $F(r)$ such that

$f(z_2, z_1) = F[f(z_1, z_2)]$. Thus, disregarding IV, we may introduce the obviously symmetric distance function

$$g(z_1, z_2) = f(z_1, z_2)+f(z_2, z_1) = G[f(z_1, z_2)],$$

where $G(r) = r+F(r)$. Also $g(z_1, z_2)$ is non-negative, and equal to zero if and only if $z_2 = z_1$. Notice that $F[F(r)] = r$.

5. Between the distance functions $d_\varepsilon(z_1, z_2)$ ($\varepsilon = -1, 1$) introduced in (12.42) and (12.8) and those of (13.31) one has the relation

$$d_\varepsilon(z_1, z_2) = 1+\varepsilon f_\varepsilon(z_1, z_2)^2. \tag{13.7}$$

6. *$\mathcal{G}$-congruence.* $\mathcal{G}$-geometry is the study of those properties of plane geometrical figures $\mathcal{F}$ (for instance, systems of $\mathcal{G}$-straight lines or segments on such lines, as $\mathcal{G}$-triangles; $\mathcal{G}$-circles, $\mathcal{G}$-cycles, etc.) which are not changed if $\mathcal{F}$ is subjected to any transformation of the group $\mathcal{G}$. Two figures $\mathcal{F}_1$, $\mathcal{F}_2$ in $\mathcal{D}$ are said to be 'congruent with respect to the group $\mathcal{G}$,' briefly, '$\mathcal{G}$-congruent,' if there is a transformation $\mathfrak{T}$ in $\mathcal{G}$ which transforms $\mathcal{F}_1$ into $\mathcal{F}_2$ (so that after the transformation, which actually affects only $\mathcal{F}_1$, both figures coincide; $\mathcal{F}_2$ is assumed to be situated in the image plane of the transformation). In symbols, we write $\mathcal{F}_1 \underset{G}{=} \mathcal{F}_2$.

This notion of $\mathcal{G}$-congruence possesses the usual four properties of equality (or equivalence) corresponding to the fundamental properties which are postulated for a general group:

1. Two figures in $\mathcal{D}$ are either $\mathcal{G}$-congruent or not. There is no third possibility.

2. Every figure $\mathcal{F}$ is $\mathcal{G}$-congruent to itself: $\mathcal{F} \underset{G}{=} \mathcal{F}$, because $\mathcal{G}$ contains the identity transformation $\mathfrak{E}$.

3. If $\mathcal{F}_1 \underset{G}{=} \mathcal{F}_2$, then $\mathcal{F}_2 \underset{G}{=} \mathcal{F}_1$. Indeed if a transformation $\mathfrak{T}$ in $\mathcal{G}$ carries $\mathcal{F}_1$ into coincidence with $\mathcal{F}_2$, then the inverse $\mathfrak{T}^{-1}$ (which is contained in $\mathcal{G}$) will take $\mathcal{F}_2$ into coincidence with $\mathcal{F}_1$.

4. If $\mathcal{F}_1 \underset{G}{=} \mathcal{F}_2$ and $\mathcal{F}_2 \underset{G}{=} \mathcal{F}_3$, then $\mathcal{F}_1 \underset{G}{=} \mathcal{F}_3$. By hypothesis there is a transformation $\mathfrak{T}_1$ in $\mathcal{G}$ which carries $\mathcal{F}_1$ into the position of $\mathcal{F}_2$, and a $\mathfrak{T}_2$ that carries $\mathcal{F}_2$ into $\mathcal{F}_3$; then the product $\mathfrak{T}_2\mathfrak{T}_1$ (which is an element of $\mathcal{G}$) will carry $\mathcal{F}_1$ into coincidence with $\mathcal{F}_3$.

It may be noted that associativity of the group multiplication has no counterpart among the properties of $\mathcal{G}$-congruence because composition of transformations is automatically associative.

§ 14. Hyperbolic geometry

In this section it will be shown that hyperbolic geometry as defined in § 13, **b** actually represents a consistent geometrical theory; it is not at all difficult to prove that this geometry satisfies the system of axioms obtained by omitting the postulate of parallels from the axioms of euclidean geometry. A detailed discussion of this aspect will not be given.

a. *Hyperbolic straight lines and distance.* $\mathcal{U}_+$-straight lines will be called hyperbolic straight lines. Those through 0 are the segments within the unit circle of ordinary straight lines through 0 and are therefore orthogonal to the circumference of the unit circle. By definition (cf. § 13, **b**) all hyperbolic lines through a point z_1 within the unit circle $\mathfrak{D}$ are obtained from those through 0 by application of a hyperbolic motion, that is, a certain element $\mathfrak{H}_{z_1}^{-1}$ of the group $\mathcal{U}_+$. They are therefore represented by the arcs within $\mathfrak{D}$ of the circles of the elliptic pencil of circles through z_1 and orthogonal to the unit circle Γ. All hyperbolic lines are then all arcs in $\mathfrak{D}$ of the circles of the hyperbolic bundle within which the hyperbolic motions are obtained as products of inversions (§ 12, **d**).

This is obviously a good reason for distinguishing these orthogonal circles by the name of *hyperbolic straight lines.* Indeed euclidean motions are products of inversions (reflections, cf. § 2, ex. 3) within the (parabolic) bundle of all euclidean straight lines. In the following, further reasons will be found to support the distinction of the orthogonal circles as hyperbolic straight lines.

An essential difference from euclidean geometry, however, is now easily discovered. Two hyperbolic straight lines may have no common point in $\mathfrak{D}$. Then it is possible that they have one common point at infinity, that is, on the horizon Γ. In this case they are said to be parallel. If they have neither a finite nor an infinite point in common, they are said to be *ultra-parallel.* The latter condition does not exist in euclidean geometry.

Any two hyperbolic lines are either parallel or not parallel. If a hyperbolic line $\mathfrak{l}_1$ is parallel to $\mathfrak{l}$, then $\mathfrak{l}$ is parallel to $\mathfrak{l}_1$. But if $\mathfrak{l}_1$ is parallel to $\mathfrak{l}$ and $\mathfrak{l}_2$ parallel to $\mathfrak{l}_1$, then $\mathfrak{l}_2$ is not necessarily parallel to $\mathfrak{l}$. For through every point in $\mathfrak{D}$ not on $\mathfrak{l}$ there are two different hyperbolic lines parallel to $\mathfrak{l}$, and three lines are parallel if and only if they have the same 'point at infinity.'

Now we shall replace the hyperbolic distance function $f_{-1}(z_1, z_2)$ of (13.31) ($\varepsilon = -1$) by another two-point invariant. Let z_1, z_2 be two distinct points in $\mathfrak{D}$. They determine a hyperbolic straight line, that is, a circle orthogonal to the unit circle Γ, which it meets at the two points ζ_1, ζ_2. We arrange that on the ortho-circle the four points are situated in the order $\zeta_2, z_1, z_2, \zeta_1$.

There is a hyperbolic motion $\mathfrak{H}$ which maps z_1 onto 0 and z_2 onto a real number $x > 0$. It maps the hyperbolic straight line through z_1 and z_2 onto the line through 0 and x, that is, the segment of the real axis within the unit circle. Thus

$$\mathfrak{H}(\zeta_1) = +1, \qquad \mathfrak{H}(\zeta_2) = -1.$$

As a Moebius transformation, $\mathfrak{H}$ will preserve the cross ratio. Thus

$$(z_1, z_2; \zeta_1, \zeta_2) = (0, x; +1, -1) = \frac{1+x}{1-x} > 1. \tag{14.1}$$

As an invariant this cross ratio must be a function of the hyperbolic distance function $f(z_1, z_2)$. Indeed

(14.2) $$f(z_1, z_2) = f(0, x) = x$$

and therefore

(14.21) $$(z_1, z_2; \zeta_1, \zeta_2) = \frac{1+f_{-1}(z_1, z_2)}{1-f_{-1}(z_1, z_2)}.$$

This cross ratio is equal to one if and only if $z_1 = z_2$.

We note the formula

(14.22) $$f_{-1}(z_1, z_2) = \frac{(z_1, z_2; \zeta_1, \zeta_2)-1}{(z_1, z_2; \zeta_1, \zeta_2)+1}.$$

For two points z_1, z_2 in $\mathscr{D}$ we introduce the *hyperbolic distance*

(14.23) $$D(z_1, z_2) = \log(z_1, z_2; \zeta_1, \zeta_2) = 2 \operatorname{arth} f_{-1}(z_1, z_2).$$[7]

The reason for the distinction of this invariant among all the other functions of $f_{-1}(z_1, z_2)$ is to be seen in the fact that it has some of the typical properties of the euclidean distance.

From the definition (14.23) and by elementary properties of the logarithm we conclude that for all z_1, z_2 in $\mathscr{D}$,

(14.24) $$D(z_1, z_2) > 0 \quad \text{if} \quad z_1 \neq z_2; \qquad D(z_1, z_1) = 0.$$

Moreover

$$D(z_1, z) \to \infty \quad \text{if} \quad z \to \zeta,$$

where ζ is a point on the horizon Γ. This justifies the agreement according to which we consider the horizon as the 'infinite' of the hyperbolic plane.

b. *The triangle inequality.* If z_1, z_2 are two points in $\mathscr{D}$ we denote by (z_1, z_2) the 'hyperbolic segment' which is defined by these two points on the hyperbolic straight line through z_1 and z_2. We call $D(z_1, z_2)$ the length of this segment.

Let z_1, z_2, z_3 be three distinct points in $\mathscr{D}$. They determine a hyperbolic triangle, that is, the figure whose sides are the three hyperbolic segments (z_2, z_3), (z_3, z_1), (z_1, z_2). An elementary theorem of euclidean geometry states that the sum of the lengths of two sides of a triangle is not smaller than the length of the third side. The same theorem is true in hyperbolic geometry:

Theorem A. *The hyperbolic distance D satisfies the so-called 'triangle inequality'*

(14.25) $$D(z_1, z_2)+D(z_2, z_3) \geqslant D(z_1, z_3)$$

[7] The function $\log r$ will be used for positive argument r only. By $\operatorname{arth} u = \operatorname{th}^{-1} u$ we denote the inverse of the hyperbolic tangent function $u = \operatorname{th} x$ ($= \tanh x$). Also the other hyperbolic functions $\operatorname{ch} x$ ($= \cosh x$) and $\operatorname{sh} x$ ($= \sinh x$) will occur later on. Their elementary properties are assumed to be known.

where the sign of equality is valid if and only if all three points lie on one and the same hyperbolic straight line and z_2 lies between z_1 and z_3.

Proof. There is a hyperbolic motion $\mathfrak{H}$ such that

$$\mathfrak{H}(z_2) = 0, \qquad \mathfrak{H}(z_1) = x > 0, \qquad \mathfrak{H}(z_3) = z.$$

As a motion does not change the distances it will be sufficient to verify that

$$D(x, 0)+D(0, z) \geqslant D(x, z)$$

or that

$$\operatorname{arth} x+\operatorname{arth} |z| \geqslant \operatorname{arth} \left|\frac{x-z}{1-xz}\right|.$$

The left-hand side equals $\operatorname{arth} \dfrac{x+|z|}{1+x|z|}$ and since $\operatorname{arth} u$ is monotonically increasing for $-1 < u < 1$ we need only show that

$$\text{(14.26)} \qquad \frac{x+|z|}{1+x|z|} \geqslant \frac{|x-z|}{|1-xz|} \quad \text{if} \quad 0 < x < 1, 0 < |z| < 1.$$

Indeed

$$\left|\frac{x-z}{1-xz}\right|^2 = \frac{1-x(z+\bar{z})+x^2|z|^2-(1-x^2)(1-|z|^2)}{1-x(z+\bar{z})+x^2|z|^2}$$

$$= 1-\frac{(1-x^2)(1-|z|^2)}{|1-xz|^2}$$

and

$$\left(\frac{x+|z|}{1+x|z|}\right)^2 = \frac{1+2x|z|+x^2|z|^2-1+x^2+|z|^2-x^2|z|^2}{1+2x|z|+x^2|z|^2}$$

$$= 1-\frac{(1-x^2)(1-|z|^2)}{(1+x|z|)^2}$$

Hence the difference $\left(\dfrac{x+|z|}{1+x|z|}\right)^2 - \left|\dfrac{x-z}{1-xz}\right|^2$ becomes

$$(1-x^2)(1-|z|^2)\left(\frac{1}{|1-xz|^2}-\frac{1}{(1+x|z|)^2}\right)$$

$$= (1-x^2)(1-|z|^2)\frac{2x(|z|+\operatorname{Re} z)}{|1-xz|^2(1+x|z|)^2} \geqslant 0$$

with the sign of equality if and only if $\operatorname{Re} z = -|z|$, that is if z lies on the negative real axis. This means that equality in (14.25) stands if and only if z_1, z_2, z_3 lie in this order on one and the same hyperbolic straight line. In this case the equation

$$\text{(14.27)} \qquad D(z_1, z_2)+D(z_2, z_3) = D(z_1, z_3)$$

expresses the fact that the hyperbolic distance is *additive*.

From Theorem A follows an important property of the hyperbolic straight line which has its analogue in euclidean geometry, namely the minimum length property.

Let $\mathscr{C}$ be an arc of a continuous curve in $\mathscr{D}$: $z = z(t)$, $(0 \leqslant t \leqslant 1)$. Let $z(0) = a$, $z(1) = b$. Consider a partition $0 = t_0 < t_1 < \ldots < t_n = 1$ and the corresponding partition of $\mathscr{C}$ by the points

$$z_\nu = z(t_\nu) \qquad (\nu = 1, 2, \ldots, n).$$

They determine a polygon inscribed in the arc $\mathscr{C}$. The sides of this polygon Π are the segments $(z_{\nu-1}, z_\nu)$. By definition

$$L(\Pi) = \sum_{\nu=1}^{n} D(z_{\nu-1}, z_\nu) \tag{14.3}$$

is the hyperbolic length of Π, and we call

$$L = \sup_{(\Pi)} L(\Pi) \tag{14.31}$$

the (possibly infinite) hyperbolic length of $\mathscr{C}$. By Theorem A, $L(\Pi)$ will not decrease if the partition Π is refined. If $\mathscr{C}$ coincides with the hyperbolic segment (a, b), then each $L(\Pi)$ will be equal to $D(a, b)$ on account of the second half of Theorem A. Thus in the case of a hyperbolic segment the new definition of length coincides with the original definition.

If, however, $\mathscr{C}$ admits of a partition such that not all the points z_ν are on the segment (a, b), then for this partition certainly $L(\Pi) > D(a, b)$. Since $L \geqslant L(\Pi)$, we have

Theorem B. *Among all the arcs $\mathscr{C}$ in $\mathscr{D}$ connecting two points a and b, the hyperbolic segment (a, b) has minimum hyperbolic length.*

c. *Hyperbolic circles and cycles.* For the definition of these curves we refer to § 13, **b.** They are all ordinary circles and arcs of circles respectively, within $\mathscr{D}$. Conversely, every circle situated entirely within $\mathscr{D}$ represents a hyperbolic circle $\mathfrak{C}$. Its hyperbolic ($\mathscr{U}_+$-) centre z_0 is obtained as the intersection of two circles, both orthogonal to $\mathfrak{C}$ and to the unit circle Γ. All the points z of $\mathfrak{C}$ have constant hyperbolic distance $D(z_0, z) = r$ from z_0; we call r the hyperbolic radius of $\mathfrak{C}$. Hyperbolically concentric circles form part of a hyperbolic pencil which contains Γ and has z_0 as one of its point circles.

All hypercycles generated by the proper hyperbolic Moebius transformation $\mathfrak{H}$ [cf. (13.1)] are arcs of circles inside $\mathscr{D}$, bounded by the fixed points γ_1, γ_2 of $\mathfrak{H}$, two points on the unit circle Γ. These circles belong to the elliptic pencil of circles through γ_1 and γ_2. Exactly one of these hypercycles is a hyperbolic straight line. Conversely every pencil of circle arcs inside $\mathscr{D}$ and bounded by two points γ_1, γ_2 on Γ represents a system of hypercycles.

The hyperbolic pencil of circles orthogonal to the pencil of hypercycles through γ_1 and γ_2 (and thus orthogonal to Γ) has γ_1 and γ_2 as point circles. The circles of this (hyperbolic) pencil are thus hyperbolic straight lines.

Theorem C. *Let $\mathfrak{h}_1, \mathfrak{h}_2$ be two hypercycles through the two distinct points γ_1, γ_2 on Γ. The hyperbolic distance between $\mathfrak{h}_1, \mathfrak{h}_2$, measured along the orthogonal hyperbolic straight lines, is constant.*

Proof. The hypercycles $\mathfrak{h}_1, \mathfrak{h}_2$ are generated by one and the same (hyperbolic) hyperbolic motion $\mathfrak{H}$. Let z_1 on $\mathfrak{h}_1$, z_2 on $\mathfrak{h}_2$ be two points on one and the same hyperbolic straight line, perpendicular to $\mathfrak{h}_1$ and $\mathfrak{h}_2$. Then $\mathfrak{H}^s(z_1)$ and $\mathfrak{H}^s(z_2)$ are two points on $\mathfrak{h}_1$ and $\mathfrak{h}_2$ respectively, again on one and the same hyperbolic straight line, orthogonal to $\mathfrak{h}_1$ and $\mathfrak{h}_2$, and for every pair of points of this kind there is a value of the real parameter s. Because of the invariance of the distance,

$$D[\mathfrak{H}^s(z_1), \mathfrak{H}^s(z_2)] = D(z_1, z_2), \qquad \text{q.e.d.} \tag{14.4}$$

The hyperbolic motion represented by the hyperbolic Moebius transformation $\mathfrak{H}$ (leaving invariant the circles of the elliptic pencil through γ_1, γ_2) is obtained as the product of two inversions within the pencil of hyperbolic straight lines orthogonal to the elliptic pencil of hypercycles.

A hyperbolic cycle is a *horocycle* if the generating transformation $\mathfrak{H}$ is parabolic. All horocycles represented by the parametric equation $z = \mathfrak{H}^s(z_0)$ for z_0 inside $\mathcal{D}$, are euclidean circles touching the unit circle Γ from inside at the (only) fixed point γ of the parabolic transformation $\mathfrak{H}$. The system of all horocycles through γ is therefore a parabolic pencil of circles.

The orthogonal pencil of a parabolic pencil is also a parabolic pencil, having the same point γ as point of common contact. This second pencil consists of circles carrying hyperbolically parallel hyperbolic straight lines. It is readily seen (cf. Theorem C) that, measured along these lines, the horocycles of the first pencil through γ have constant hyperbolic distance from one another.

The (parabolic) hyperbolic motion $\mathfrak{H}$ is the product of two inversions within the orthogonal parabolic pencil of hyperbolically parallel hyperbolic straight lines.

d. *Hyperbolic trigonometry.* Let z_1, z_2, z_3 be the vertices of a hyperbolic triangle; its sides are the hyperbolic segments (z_1, z_2), (z_1, z_3), (z_2, z_3) and

$$D(z_1, z_2) = a, \qquad D(z_1, z_3) = b, \qquad D(z_2, z_3) = c$$

are their hyperbolic lengths. First let the triangle be 'right' or rectangular, having the right angle at the vertex z_1. As in the case of euclidean geometry we shall find the whole of hyperbolic trigonometry depending on the relation between the catheti a and b and the hypotenuse c. Thus we derive first the 'hyperbolic Pythagoras.'

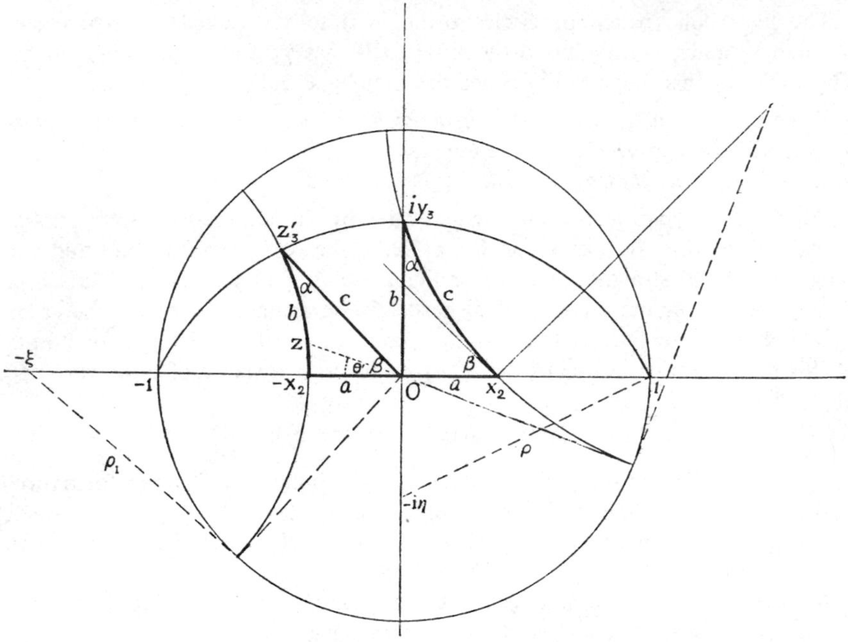

FIG. 24

For this purpose we apply a hyperbolic motion by which the triangle is taken into a 'normal position' such that the catheti coincide with the real and imaginary axes and the hypotenuse lies in the first quadrant of the coordinate system, as shown in Figure 24. We have then

(14.5)
$$\begin{cases} a = D(0, x_2) = 2 \operatorname{arth} x_2 & \text{or} \quad x_2 = \operatorname{th} \dfrac{a}{2} > 0 \\ b = D(0, iy_3) = 2 \operatorname{arth} y_3 & \text{or} \quad y_3 = \operatorname{th} \dfrac{b}{2} > 0 \\ c = D(x_2, iy_3) = 2 \operatorname{arth} \left| \dfrac{x_2 - iy_3}{1 + ix_2 y_3} \right| & \end{cases}$$

and therefore[8]

(14.51)
$$\operatorname{th}^2 \frac{c}{2} = \frac{x_2{}^2 + y_3{}^2}{1 + x_2{}^2 y_3{}^2} = \frac{\operatorname{th}^2 (a/2) + \operatorname{th}^2 (b/2)}{1 + \operatorname{th}^2 (a/2) \operatorname{th}^2 (b/2)}.$$

[8] Following the general custom we denote by $\operatorname{th}^2 x$ the function which properly should be denoted by $(\operatorname{th} x)^2$.

To simplify this formula observe that

$$\mathrm{th}^2\frac{x}{2} = \frac{\mathrm{ch}x-1}{\mathrm{ch}x+1} \qquad (x = a, b, c).$$

Hereby (14.51) is changed into

$$\frac{\mathrm{ch}c-1}{\mathrm{ch}c+1} = \frac{\mathrm{ch}a\,\mathrm{ch}b-1}{\mathrm{ch}a\,\mathrm{ch}b+1},$$

whence

$$\mathrm{ch}c = \mathrm{ch}a\,\mathrm{ch}b. \tag{14.52}$$

Remark. This form of the 'hyperbolic Pythagoras' shows that hyperbolic geometry is 'locally euclidean' which means that for 'small' right triangles the euclidean Pythagoras is approximately true. In fact, using for ch x the well-known power series, viz. $\mathrm{ch}x = 1+\frac{1}{2}x^2+\dots$, and taking into account the quadratic terms only, we conclude from (14.52) that for small a, b, c, $c^2 = a^2+b^2$ is approximately true.

Now we shall derive the relations that exist between the sides and the angles of the right triangle (z_1, z_2, z_3). Let β be the angle at z_2 and α the angle at z_3. We choose another normal position for the triangle, namely so that z_2 is taken (hyperbolically) into the origin and the cathetus a coincides with the negative real axis. This new position of the triangle is obtained from the previous one by the following construction (cf. Figure 24). Draw the hypercycle through $+1$ and -1 which has the constant hyperbolic distance b from the real axis; let ρ be its euclidean radius and $-i\eta$ its centre; then

$$\eta^2 = \rho^2-1, \qquad \eta = \rho-y_3$$

$$\eta = \tfrac{1}{2}\left(\frac{1}{y_3}-y_3\right) > 0. \tag{14.53}$$

Hence the equation of the hypercycle is

$$x^2+(y+\eta)^2 = \rho^2$$

or

$$z\bar{z}+2\eta\,\mathrm{Im}\,z = \rho^2-\eta^2 = 1. \tag{14.54}$$

The hypotenuse of our triangle lies now on the hyperbolic straight line (= ordinary line) through 0:

$$z = re^{i\beta'} \qquad (\beta' = \pi-\beta, 0 \leqslant r \leqslant r_3). \tag{14.541}$$

It meets the hypercycle at the vertex z_3' of our triangle in the new position, $|z_3'| = r_3$. Thus by (14.54)

$$r_3^2+2\eta r_3 \sin\beta = 1,$$

whence by (14.53)

(14.542) $$\frac{y_3}{1-y_3^2} = \frac{r_3}{1-r_3^2} \sin \beta.$$

According to the construction of the triangle in the new position, we have (cf. Figure 24):

(14.55) $$x_2 = \operatorname{th} \frac{a}{2}, \qquad r_3 = \operatorname{th} \frac{c}{2}, \qquad \left| \frac{z_3'+x_2}{1+x_2 z_3'} \right| = \operatorname{th} \frac{b}{2} = y_3.$$

Thus (14.542) becomes a relation between $\sin \beta$ and hyperbolic tangents; writing the latter in terms of ch and sh we get

$$\operatorname{ch} \frac{b}{2} \operatorname{sh} \frac{b}{2} = \operatorname{ch} \frac{c}{2} \operatorname{sh} \frac{c}{2} \sin \beta$$

or

(14.56) $$\operatorname{sh} b = \operatorname{sh} c \sin \beta, \qquad \operatorname{sh} a = \operatorname{sh} c \sin \alpha,$$

where the second formula is an obvious consequence of the first one.

In a similar manner, calculating the point z_3' as intersection of the line (15.541) (still referring to the second normal position of the triangle) with the ortho-circle through the point $-x_2$, we get a relation between a, c, β (or b, c, α). Let $-\xi$ be the centre of the ortho-circle; we find

$$\xi = \tfrac{1}{2}\left(\frac{1}{x_2}+x_2\right)$$

and, for r_3, the condition

$$r_3^2+2\xi r_3 \cos \beta' = r_3^2-2\xi r_3 \cos \beta = -1.$$

This may be written in the form

$$\frac{x_2^2+1}{x_2} \cos \beta = \frac{r_3^2+1}{r_3}$$

whence, by (14.55)

(14.57) $$\operatorname{th} c \cos \beta = \operatorname{th} a, \qquad \operatorname{th} c \cos \alpha = \operatorname{th} b.$$

Now we have the basis for the trigonometry of an arbitrary hyperbolic triangle. We assume it to be in a normal position as indicated in Figure 25. The side a lies on the real axis and the corresponding height $h = h_a$ on the imaginary axis. Let a_1, a_2 be the two hyperbolic segments (as well as their hyperbolic lengths) into which the foot of this height divides the side a; then

$$a_1 = \begin{cases} a-a_2 & \text{if the height lies inside the triangle (Figure 25a)} \\ a+a_2 & \text{if the height lies outside the triangle (Figure 25b).} \end{cases}$$

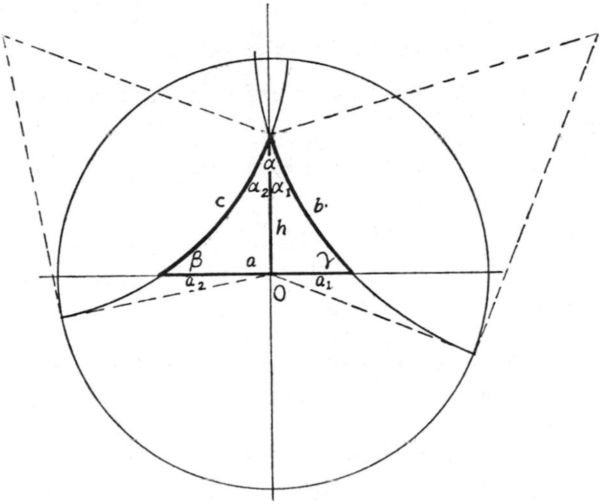

FIG. 25a

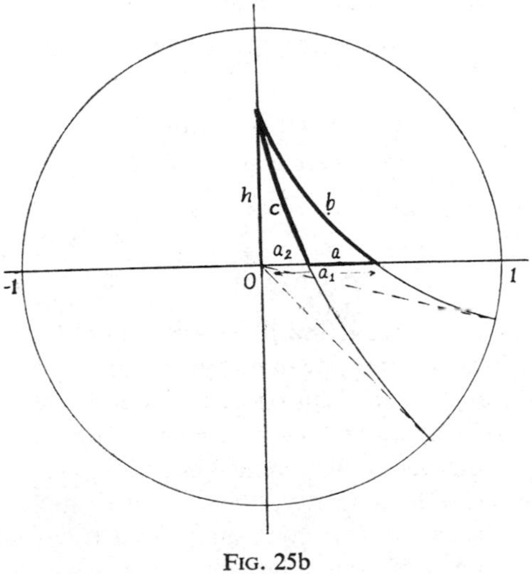

FIG. 25b

By (14.57) and (14.52)

$$\operatorname{th} a_1 = \pm \operatorname{th} b \cos \gamma, \qquad \operatorname{ch} h = \frac{\operatorname{ch} b}{\operatorname{ch} a_1},$$

and therefore

$$\begin{aligned}\operatorname{ch} c = \operatorname{ch} a_2 \operatorname{ch} h &= \operatorname{ch}(a \mp a_1) \operatorname{ch} h \\ &= (\operatorname{ch} a \mp \operatorname{sh} a \operatorname{th} a_1) \operatorname{ch} b.\end{aligned}$$

Hence

(14.58) $$\text{ch}c = \text{ch}a\ \text{ch}b - \text{sh}a\ \text{sh}b \cos\gamma,$$

This is the analogue of the 'cosine theorem' of elementary trigonometry.

Further, by (14.56),

$$\text{sh}h = \text{sh}b \sin\gamma = \text{sh}c \sin\beta,$$

whence

(14.59) $$\frac{\text{sh}a}{\sin\alpha} = \frac{\text{sh}b}{\sin\beta} = \frac{\text{sh}c}{\sin\gamma}.$$

This is the so-called sine theorem.

e. *Applications.* The trigonometric formulae of § 14, **d** enable us to derive some important facts that show the fundamental difference between euclidean and hyperbolic geometry.

According to a well-known theorem of euclidean geometry, the sum of the three angles in every (euclidean) triangle is equal to π. This is no longer true in hyperbolic geometry.

Theorem D. *The sum of the angles in every hyperbolic triangle is less than* π.

Proof. Evidently it is sufficient to prove the theorem for a right triangle. Then, by (14.57) and (14.56),

$$\cos(\alpha+\beta) = \frac{\text{th}b}{\text{th}c}\frac{\text{th}a}{\text{th}c} - \frac{\text{sh}a}{\text{sh}c}\frac{\text{sh}b}{\text{sh}c} = \frac{\text{sh}a\ \text{sh}b}{\text{sh}^2 c}(\text{ch}c-1)$$
$$= \sin\alpha \sin\beta(\text{ch}c-1) > 0;$$

therefore $$\alpha+\beta < \frac{\pi}{2},$$

which proves the theorem since α and β are assumed to be positive.

Given three lengths a, b, c (euclidean or hyperbolic) which satisfy the triangle inequality, then a triangle (euclidean or hyperbolic) can be constructed whose sides have just these lengths, and every triangle with the same measures is congruent (in the euclidean or hyperbolic sense ($\mathscr{U}_+$-congruent) respectively) to the first one. In euclidean geometry, however, three angles α, β, γ whose sum equals π are not sufficient to determine uniquely a triangle whose angles are α, β, γ; there is an infinity of non-congruent triangles whose angles have these values in the same order. All these triangles are said to be *similar* to one another. The following theorem shows that *there is no similarity in hyperbolic geometry.*

Theorem E. *If* α, β, γ *are three angles that satisfy the condition* $\alpha+\beta+\gamma < \pi$, *then there is a hyperbolic triangle having these angles, and any two such triangles are hyperbolically congruent.*[9]

[9] This means '$\mathscr{U}_+$-congruent,' cf. § 13, ex. 6.

Proof. First, suppose that $\gamma = \pi/2$; hence $\alpha+\beta < \pi/2$. Using the standard notation of Figure 24, we have by (14.56) and (14.57) the relation

$$\frac{\text{sh}a}{\text{th}a} = \frac{\text{sh}c}{\text{th}c}\frac{\sin\alpha}{\cos\beta}, \quad \text{or} \quad \text{ch}a = \text{ch}c\frac{\sin\alpha}{\cos\beta},$$

whence by (14.52)

$$\text{(14.6)} \qquad \text{ch}b = \frac{\cos\beta}{\sin\alpha}, \qquad \text{ch}a = \frac{\cos\alpha}{\sin\beta}, \qquad \text{ch}c = \cot\alpha\cot\beta.$$

The expressions on the right-hand sides of these three relations are greater than 1; indeed, by hypothesis, $\alpha < (\pi/2)-\beta$, and thus $\cos\alpha > \cos[(\pi/2)-\beta] = \sin\beta$, and likewise $\cos\beta > \sin\alpha$. Hence the sides of a right triangle are uniquely defined by the two angles α, β. Indeed (14.52) is satisfied and therefore by (14.58) $\gamma = \pi/2$. Thus there is a right triangle which has a, b as sides, c as hypotenuse, and the angles α, β.

Now let α, β, γ be three prescribed angles that satisfy the condition $\alpha+\beta+\gamma < \pi$. We allow α to be obtuse or a right-angle, but β and γ must be acute. Let $\alpha = \alpha_1+\alpha_2$ (cf. Figure 25a). We wish to determine α_1 so that the two right triangles with the angles γ, α_1, $\pi/2$ and β, α_2, $\pi/2$ have one cathetus in common which will become the height $h_a = h$ of the triangle with the angles α, β, γ. By (14.6),

$$\text{(14.61)} \qquad \text{ch}h = \frac{\cos\gamma}{\sin\alpha_1} = \frac{\cos\beta}{\sin\alpha_2}.$$

Equality of the two quotients means that

$$\text{(14.62)} \qquad \frac{\sin\alpha_2}{\sin\alpha_1} = \sin\alpha\cot\alpha_1 - \cos\alpha = \frac{\cos\beta}{\cos\gamma},$$

whence we obtain $\cot\alpha_1$, and thereby $1/\sin^2\alpha_1$, in terms of α, β, and γ.

But for the existence of a real h it is necessary and sufficient that $\text{ch}\,h > 1$. Assuming that there is an h that satisfies this condition we have

$$\text{(14.63)} \qquad \alpha_1+\gamma < \frac{\pi}{2}$$

which by (14.62) is equivalent to

$$\tan\gamma < \cot\alpha_1 = \frac{1}{\sin\alpha}\left(\frac{\cos\beta}{\cos\gamma}+\cos\alpha\right)$$

or

$$\text{(14.64)} \qquad \cos(\alpha+\gamma)+\cos\beta > 0.$$

Now by hypothesis $0 < \beta < \pi/2$, and therefore $\cos\beta > 0$. If also $\alpha+\gamma \leqslant \pi/2$, the inequality (14.64) is satisfied. If $\alpha+\gamma > \pi/2$, then by (14.64)

$$\cos\beta > -\cos(\alpha+\gamma) = \cos[\pi-(\alpha+\gamma)]$$

which implies that $\beta < \pi-\alpha-\gamma$ and conversely. Thus (14.63) is established in all cases.

From it we conclude that $\cot\alpha_1 > \tan\gamma$ or

$$\sin^2\alpha_1 = \frac{1}{1+\cot^2\alpha_1} < \frac{1}{1+\tan^2\gamma} = \cos^2\gamma,$$

so that indeed (14.61) can be satisfied by a real h.

Finally, by (14.6),

$$\mathrm{ch}\,a_1 = \frac{\cos\alpha_1}{\sin\gamma}, \qquad \mathrm{ch}\,a_2 = \frac{\cos\alpha_2}{\sin\beta}, \qquad a = a_1+a_2,$$

and by (14.52) and (14.61),

$$\mathrm{ch}\,b = \mathrm{ch}\,a_1\,\mathrm{ch}\,h = \mathrm{ch}\,a_1\frac{\cos\gamma}{\sin\alpha_1} = \cot\alpha_1\cot\gamma \tag{14.65}$$

$$\mathrm{ch}\,c = \mathrm{ch}\,a_2\,\mathrm{ch}\,h = \mathrm{ch}\,a_2\frac{\cos\beta}{\sin\alpha_2} = \cot\alpha_2\cot\beta.$$

Thus the triangle with the prescribed angles exists and is unique up to congruence.

Writing (14.65) in the form

$$\mathrm{ch}\,b\sin\gamma\sin\alpha = \cot\alpha_1\cos\gamma\sin\alpha,$$

and using the second equation (14.62), the right-hand expression takes the form

$$\left(\frac{\cos\alpha}{\sin\alpha}+\frac{\cos\beta}{\cos\gamma\sin\alpha}\right)\cos\gamma\sin\alpha = \cos\alpha\cos\gamma+\cos\beta.$$

Thus we get the 'dual cosine theorem' [cf. (14.58)]:

$$\cos\beta = -\cos\alpha\cos\gamma+\sin\alpha\sin\gamma\,\mathrm{ch}\,b. \tag{14.66}$$

Examples

1. In the preceding discussion the interior of the unit circle has been chosen as the hyperbolic plane. The interior of any other circle would be equally suitable for the purpose, in particular we can choose a straight line, for example, the real axis, as the horizon. The hyperbolic plane $\mathscr{D}$ will then be the upper half plane $\mathrm{Im}\,z > 0$. The group of motions in this $\mathscr{G}$-geometry will be the group of all real Moebius transformations (cf. § 8, **d**) with positive determinant. Hyperbolic straight lines are in this case the semi-circles orthogonal to the real axis. The distance function $f(z_1, z_2)$ may be found by means of the device of

§ 13, ex. 3. Choose $z_0 = i$ and let $\mathfrak{T}$ be a Moebius transformation that maps the upper half plane onto the interior of the unit circle so that $\mathfrak{T}(i) = 0$; let

$\mathfrak{T} = \begin{pmatrix} i & 1 \\ 1 & i \end{pmatrix}$ and $\mathfrak{H}_{z_2}(z) = (z-x_2)/y_2$ if $z_2 = x_2+iy_2$. Then

$$f(z_1, z_2) = \left| \frac{z_1-z_2}{z_1-\bar{z}_2} \right| = (z_1, \bar{z}_1; z_2, \bar{z}_2)^{\frac{1}{2}}.$$

For all geometrical purposes this 'half-plane model' of hyperbolic geometry is entirely equivalent to the 'unit circle model.'[10] We have preferred the latter on account of its applications in the theory of analytic functions.

2. There are other models of hyperbolic geometry, that is, representations by means of elements of euclidean geometry. Hyperbolic geometry in the interior of the unit circle may be transplanted onto the upper hemisphere by means of stereographic projection. The equator then plays the role of the horizon and hyperbolic straight lines are represented by the spherical semi-circles orthogonal to the equator; their planes are perpendicular to the plane of the equator. An algebraic representation of the motion group has been established in § 12, ex. 4.

By a further projection, namely a perspectivity from the infinite point of the ζ-axis as centre, we may return into the interior of the unit circle. It is readily seen that by this parallel projection the spherical semi-circles orthogonal to the equator are mapped onto the euclidean segments within the unit circle. Thus we arrive at another model of hyperbolic geometry where the straight lines are represented by chords of the unit circle. The group of hyperbolic motions in this model is represented by the group of all those projective transformations in the plane that leave the unit circle invariant; these transformations are given by (12.75).

Since the parallel projection mapping the upper hemisphere onto the interior of the unit circle is not conformal except at the North pole, it is evident that in general the measure of hyperbolic angles in the second unit-circle model is different from the measure of the corresponding euclidean angles. For an exposition of hyperbolic geometry on this model we may refer to the book by Baldus [1].

In all the further examples we refer to the 'conformal unit circle model' (also called the Poincaré circle model) as dealt with in the text of § 14, except § 15, ex. 5.

3. An inversion with respect to an ortho-circle in $\mathscr{D}$, that is, a hyperbolic straight line, will be called a *hyperbolic symmetry*. The main properties of these symmetries are the following. A hyperbolic symmetry transforms the interior

[10] The half-plane model is often referred to as the Poincaré model.

of the unit circle onto itself and it leaves the unit circle (as a curve) invariant. For every point z within the unit circle ($|z| < 1$) the hyperbolically symmetric point z^* also lies inside the unit circle (cf. Figure 26). A hyperbolic symmetry maps hyperbolic straight lines onto hyperbolic straight lines. The hyperbolic distance is invariant with respect to hyperbolic symmetries.

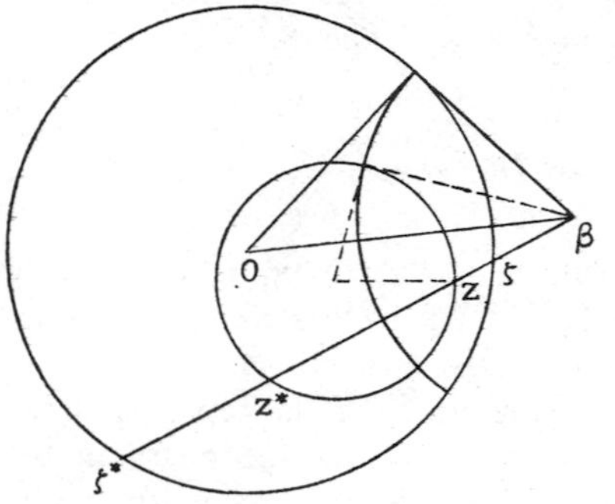

FIG. 26

4. For a given hyperbolic segment (z_1, z_2) in $\mathscr{D}$, find by geometrical construction the congruent segment in normal position, that is, so that 0 is one endpoint while the other endpoint lies on the positive real axis. Likewise for a given triangle find the congruent triangle in normal position (Figure 24).

5. Prove the uniqueness, up to a constant positive factor, of the hyperbolic distance $D(z_1, z_2)$ if it is required to be

1. invariant by hyperbolic motion,
2. non-negative,
3. additive on one hyperbolic straight line [cf. (14.27)].

Invariance implies that $D(z_1, z_2) = \phi[f_{-1}(z_1, z_2)]$ where $\phi(u)$ is a real non-negative function for $0 \leqslant u < 1$. Additivity of D implies for $\phi(u)$ the functional equation

$$\phi\left(\frac{x_1 - x_2}{1 - x_1 x_2}\right) = \phi(x_1) - \phi(x_2), \qquad (x_1 \geqslant x_2),$$

whence it is evident that $\phi(u)$ is monotonic and $\phi(0) = 0$. It is also continuous. Indeed let $\lim_{u \to +0} \phi(u) = \alpha$; then $\alpha \geqslant 0$. From the functional equation it follows that

$$\lim_{x_1 \to x_2 + 0} \phi(x_1) = \phi(x_2) + \alpha,$$

and since a monotonic function can have only a countable number of discontinuities, $\alpha = 0$, that is, $\phi(u)$ is continuous from the right. The functional equation implies that it is also continuous from the left and therefore continuous.

Now one solution of the equation is known: $\phi(x) = \operatorname{arth} x$. Let $\psi(x)$ be another solution; then

$$\psi^{-1}[\psi(x_1) - \psi(x_2)] = \phi^{-1}[\phi(x_1) - \phi(x_2)].$$

Thus putting $\phi(x_i) = X_i$ and $\psi[\phi^{-1}(X)] = F(X)$ we have for $F(X)$ the well-known functional equation

$$F(X_1) - F(X_2) = F(X_1 - X_2)$$

every continuous solution of which is given by $F(X) = cX$, where c is constant. Thus $\psi(x) = c\phi(x)$. (Cf. § 15, ex. 5.)

6. *Angle of parallelism.* In the right triangle with the vertices 0, x_2, iy_3 (cf. Figure 24) let x_2 tend to $+1$, that is, (hyperbolically) one of the points at infinity of the hyperbolic straight line which carries the cathetus a. The triangle will degenerate into a so-called 'simply asymptotic triangle' which has the angle $\beta = 0$ and the hypotenuse c parallel to the cathetus a. We observe that the angle α of this simply asymptotic triangle will not be a right angle as we should expect it to be in euclidean geometry where a line parallel to a given line $\mathfrak{l}$ forms a right angle with a perpendicular on $\mathfrak{l}$. In hyperbolic geometry two parallel lines do not possess a common perpendicular at all. In the process of degeneration of our triangle, $a \to \infty$, also $c \to \infty$, and therefore th$c \to 1$; thus by (14.57) the angle α approaches a limit angle α^* such that

$$\cos \alpha^* = \text{th} b.$$

Hence $\alpha^* < \pi/2$ and α^* will be the smaller, the greater we choose b. The angle α^* is called the *angle of parallelism.*

7. All hyperbolic straight lines passing through a point z_0 of $\mathscr{D}$ which are ultra-parallel to a given hyperbolic straight line $\mathfrak{l}$ not through z_0, are situated within the external angle formed by the two hyperbolic parallels to $\mathfrak{l}$ through z_0. Show that any two non-intersecting ultra-parallels to $\mathfrak{l}$ have a common perpendicular hyperbolic line.

8. Construct an asymptotic triangle having the angle sum zero. Show that four common hyperbolic parallels exist for two non-parallel hyperbolic lines.

9. The points of the completed plane outside the unit circle are called the 'ultra-infinite points' of hyperbolic geometry. Although they are alien elements in this geometry, they are sometimes usefully employed, for instance in the definition of a distance function as indicated in (12.42) (cf. § 13, ex. 5).

10. The circle about 0 which is hyperbolically congruent to a hyperbolic circle with the hyperbolic radius r, has the euclidean radius $\rho = \text{th}\,\frac{1}{2}r$.

11. *Area in hyperbolic geometry.* Let $\mathscr{A}$ be a closed region in $\mathscr{D}$, surrounded by one, or more than one, simply closed curve. As area $m(\mathscr{A})$ of $\mathscr{A}$ we introduce a non-negative number which has the following properties. (i) As a function of $\mathscr{A}$ the area $m(\mathscr{A})$ is *additive*, that is, if $\mathscr{A}$ consists of two non-overlapping parts $\mathscr{A}_1$, $\mathscr{A}_2$, then $m(\mathscr{A}) = m(\mathscr{A}_1)+m(\mathscr{A}_2)$. (ii) The area is *invariant*, that is, if $\tilde{\mathscr{A}}$ is the image of $\mathscr{A}$ by a hyperbolic motion or symmetry (cf. ex. 3), then $m(\tilde{\mathscr{A}}) = m(\mathscr{A})$.

We shall define $m(\mathscr{A})$ by a double integral taken over the region $\mathscr{A}$, thus assuring the additivity. To formulate the invariance condition we apply the transformation formula for double integrals.[11] For this purpose we need the

[11] Cf. Courant [1], vol. II, p. 253.

functional determinant (Jacobian) of a Moebius transformation $Z = \mathfrak{H}(z) = (az+b)/(cz+d)$. Putting $z = x+iy$, $Z = X+iY$, this determinant will be

$$(14.7) \qquad \Delta(z) = \frac{\partial X}{\partial x}\frac{\partial Y}{\partial y} - \frac{\partial X}{\partial y}\frac{\partial Y}{\partial x} = \left|\frac{d\mathfrak{H}(z)}{dz}\right|^2 = \left|\frac{ad-bc}{(cz+d)^2}\right|^2.$$ [12]

If $Z = \mathfrak{H}_{z_1}(z) = (z-z_1)/(1-\bar{z}_1 z)$ this determinant will be

$$(14.71) \qquad \Delta_{z_1}(z) = \frac{(1-|z_1|^2)^2}{|1-\bar{z}_1 z|^4}.$$

Now let $\psi(z)$ be a real function, defined in the interior of the unit circle, positive, continuous, and differentiable with respect to the two real variables x, y. Consider

$$(14.72) \qquad m(\mathscr{A}) = \iint_{\mathscr{A}} \psi(z)\, dx\, dy.$$

Invariance with respect to the hyperbolic motion $Z = \mathfrak{H}_{z_1}(z)$ is expressed in the following way:

$$m(\bar{\mathscr{A}}) = \iint_{\bar{\mathscr{A}}} \psi(Z)\, dX\, dY = \iint_{\mathscr{A}} \psi[\mathfrak{H}_{z_1}(z)]\Delta_{z_1}(z)\, dx\, dy = \iint_{\mathscr{A}} \psi(z)\, dx\, dy.$$

By contracting $\mathscr{A}$ into a single point in $\mathscr{D}$ and using the process of 'space differentiation'[13] based on the mean value theorem for double integrals, we derive from the integral condition the following functional equation for the unknown function $\psi(z)$:

$$(14.73) \qquad \psi[\mathfrak{H}_{z_1}(z)]\cdot\Delta_{z_1}(z) = \psi(z).$$

Let x and x_1 be real. We consider x_1 as parameter of the family of hyperbolic motions on the real axis

$$X = \frac{x-x_1}{1-x_1 x}$$

for which the invariance condition (14.73) becomes

$$\psi(X)(1-x_1^2) = (1-x_1 x)^4 \psi(x).$$

This is an identity in the parameter x_1. By differentiating it with respect to x_1 and putting $x_1 = 0$ we obtain the following ordinary linear differential equation for the function $\psi(x)$:

$$(x^2-1)\psi'(x) = -4x\psi(x).$$

[12] This formula is an immediate consequence of the Cauchy-Riemann differential equations which characterize X and Y as real and imaginary parts of a differentiable function of the complex variable z. Cf. Courant [1], vol. II, p. 532.

[13] Cf. Courant [1], vol. II, p. 234.

Hence

$$\psi(x) = \frac{k}{(1-x^2)^2}, \tag{14.74}$$

with an arbitrary real positive constant k, and

$$\psi(z) = \psi(|z|) = \frac{k}{(1-|z|^2)^2}. \tag{14.75}$$

Thus by (14.72)

$$m(\mathscr{A}) = k\iint_{\mathscr{A}} \frac{dx\,dy}{(1-|z|^2)^2} = k\iint_{\mathscr{A}} \frac{\rho\,d\rho\,d\theta}{(1-\rho^2)^2}, \tag{14.76}$$

using polar coordinates ρ, θ in the second integral.

If $\mathscr{A}$ is a circular disk within $\mathscr{D}$, r its hyperbolic radius and ρ_0 the euclidean radius of the hyperbolically congruent disk that has 0 as centre, then the second integral (14.76) can easily be evaluated:

$$m(\mathscr{A}) = 2\pi k\int_0^{\rho_0} \frac{\rho\,d\rho}{(1-\rho^2)^2} = \pi k\frac{\rho_0^2}{1-\rho_0^2} = \pi k \operatorname{sh}^2 \tfrac{1}{2}r \qquad \text{(cf. ex. 14).}$$

In order to have this formula so that for small values of the hyperbolic radius r it coincides approximately with the corresponding euclidean formula for the circle area, we have to take

$$k = 4. \tag{14.77}$$

[Cf. the remark under formula (14.52)].

12. *Area of the hyperbolic triangle.* Let $\mathscr{A}$ be the right triangle in the second normal position shown in Figure 24, the angle β being at the vertex 0. Let z be a point on the side b of the triangle;

$$z = \rho(\theta)e^{i\theta} \qquad (\pi-\beta \leqslant \theta \leqslant \pi)$$

and

$$s = s(\theta) = D(0, z), \qquad \rho = \operatorname{th}\frac{s}{2}.$$

From (14.76) and (14.77),

$$m(\mathscr{A}) = 4\iint_{\mathscr{A}} \frac{\rho\,d\rho\,d\theta}{(1-\rho^2)^2} = 2\int_{\theta=\pi-\beta}^{\pi}\int_{\sigma=0}^{s(\theta)} \operatorname{sh}\frac{\sigma}{2}\operatorname{ch}\frac{\sigma}{2}\,d\sigma\,d\theta$$

$$= \int_{\pi-\beta}^{\pi} [\operatorname{ch} s(\theta)-1]\,d\theta.$$

Using (14.57) in the right triangle with the vertices 0, $-x_2$, z, we have

$$\operatorname{th} s = \frac{\operatorname{th} a}{\cos\theta'}, \qquad \operatorname{ch}^2 s = \frac{\cos^2\theta}{\cos^2\theta - \operatorname{th}^2 a}, \qquad \theta' = \pi-\theta,$$

and therefore

$$m(\mathscr{A}) = -\int_{\pi-\beta}^{\pi} \frac{\cos\theta\, d\theta}{\sqrt{(\cos^2\theta - \operatorname{th}^2 a)}} - \beta = \int_0^{\sin\beta} \frac{du}{\sqrt{(1-\operatorname{th}^2 a - u^2)}} - \beta$$

$$= \operatorname{arc\,sin}(\operatorname{ch} a \sin\beta) - \beta = \operatorname{arc\,sin}(\cos\alpha) - \beta$$

$$= \frac{\pi}{2} - \alpha - \beta = \pi - \frac{\pi}{2} - \alpha - \beta.$$

The general triangle appears as the 'sum' of two right triangles as shown in Figure 25; its area is therefore the sum of the areas of the two right triangles, viz.

$$\frac{\pi}{2} - \beta - \alpha_1 + \frac{\pi}{2} - \gamma - \alpha_2 = \pi - \alpha - \beta - \gamma = \Delta_A. \tag{14.78}$$

This number Δ_A which, according to Theorem D is always positive, is known as the defect of the hyperbolic triangle with the angles α, β, γ.

Theorem F. *The hyperbolic area of a hyperbolic triangle is equal to its defect:* $m(\mathscr{A}) = \Delta_A$.

Without resorting to the preceding proof it is readily seen that the defect Δ_A of a triangle possesses the two typical properties (i) and (ii) of an area (cf. ex. 11):

(i) Additivity. The defect of a triangle is equal to the sum of the defects of the two triangles into which the given triangle is decomposed by a transversal line.

(ii) Invariance. Two congruent triangles have the same defect.

13. The hyperbolic length of an arc $\mathscr{C}$ has been introduced in (14.31). Assuming $\mathscr{C}$ to be rectifiable, we shall now derive a representation of L_C by a line integral.

By (14.23),

$$D(z_{\nu-1}, z_\nu) = 2 \operatorname{arth} f_{-1}(z_{\nu-1}, z_\nu),$$

and thus by the mean value theorem of differential calculus,

$$D(z_{\nu-1}, z_\nu) = \frac{2}{1-u_\nu^2} f_{-1}(z_{\nu-1}, z_\nu),$$

where $0 < u_\nu < f_{-1}(z_{\nu-1}, z_\nu)$. Since u_ν is smaller than any given positive number if the segment $(z_{\nu-1}, z_\nu)$ is sufficiently small, we have by (13.31) and (14.3)

$$L(\Pi) = \sum_{\nu=1}^{n} \frac{2}{1-u_\nu^2} \frac{|z_\nu - z_{\nu-1}|}{|1-\bar{z}_{\nu-1} z_\nu|} = \frac{2}{1-v_n^2} \sum_{\nu=1}^{n} \frac{|z_\nu - z_{\nu-1}|}{|1-\bar{z}_{\nu-1} z_\nu|},$$

where v_n is a certain mean value of the n numbers $u_1, \ldots, u_n$ which depends on $\mathscr{C}$ and on Π. A limit process well known from the definition of a line integral yields

$$L = \int_{\mathscr{C}} \frac{2|dz|}{1-|z|^2}.$$

14. The hyperbolic length of the circumference of a circle $\mathscr{C}$ with the hyperbolic radius r is equal to

$$L = 2\pi \operatorname{sh} r.$$

§ 15. Spherical and elliptic geometry

Spherical geometry has been defined in § 13, **b**, as the $\mathscr{G}$-geometry of the group $\mathscr{G} = \mathscr{R}$. Its plane $\mathscr{D}$ is the completed plane of the complex numbers. By stereographic projection it corresponds to a geometry on the surface of the sphere. Its interpretation in the plane leads to certain difficulties; these are overcome by the introduction of elliptic geometry.

a. *Spherical straight lines and distance.* According to the definition in § 13, **b**, $\mathscr{R}$-straight lines—from now on called 'spherical straight lines'—are circles orthogonal to the imaginary unit circle. By § 12, **c** (i), they are the stereographic images of the great circles on the sphere, analytically characterized by (3.43): With a point z they contain its antipodal point $-1/\bar{z}$; conversely, a circle that contains a pair of antipodal points is a spherical straight line. The spherical lines through 0, corresponding to the great circles through **N** and **S** on the sphere, are the ordinary straight lines through 0. Any two spherical straight lines intersect in a pair of antipodal points. Thus *there are no parallel straight lines in spherical geometry.*

As in real projective geometry there is no real point at infinity in spherical geometry; the point ∞, like all the other points of the completed plane, as image of the point **S** of the sphere, is not distinguished from the other points in the complex plane. The imaginary unit circle may be considered as an algebraic substitute for the 'absolute' or 'infinity' in spherical geometry.

After these preliminaries we turn to the definition of the spherical distance of two points z_1, z_2 in the spherical plane. Let $\mathbf{P}_1$, $\mathbf{P}_2$ be the stereographic images of z_1, z_2 on the unit sphere. Their euclidean distance, measured along the great circle through $\mathbf{P}_1$, $\mathbf{P}_2$, is evidently a two-point invariant with respect to the rotations of the sphere. If $\mathbf{P}_1$ and $\mathbf{P}_2$ are not coincident or diametrically opposite, then two different values offer themselves as distance between $\mathbf{P}_1$ and $\mathbf{P}_2$, namely the magnitude ω_{12} $(0 \leqslant \omega_{12} < \pi)$ of the angle made by the two vectors $\mathbf{OP}_1$ and $\mathbf{OP}_2$, and the value $2\pi - \omega_{12}$. A general decision as to which of the two values should be chosen as the distance $\tilde{D}(z_1, z_2)$ cannot be given.

Rather it is advisable to admit not only those two, but also all the values $\omega_{12} + 2k\pi$ and $-\omega_{12} + 2k\pi$ with an arbitrary integer k, as values of $\tilde{D}(z_1, z_2)$. Then $\tilde{D}(z_1, z_2) = 2k\pi$ indicates that $z_1 = z_2$, and conversely. For all these values of k it is found that

$$\cos \tilde{D}(z_1, z_2) \quad \text{and also} \quad \left|\tan \tfrac{1}{2}\tilde{D}(z_1, z_2)\right|$$

have the same values respectively.

By (13.31) the spherical distance function is given by

$$f_1(z_1, z_2) = \left| \frac{z_1 - z_2}{1 + \bar{z}_1 z_2} \right|.$$

If $\mathbf{P}_1$ and $\mathbf{P}_2$ have the coordinates ξ_1, η_1, ζ_1 and ξ_2, η_2, ζ_2 then

$$z_1 = \frac{\xi_1 + i\eta_1}{1 + \zeta_1}, \qquad z_2 = \frac{\xi_2 + i\eta_2}{1 + \zeta_2}.$$

By a simple calculation we find

$$f_1(z_1, z_2)^2 = \frac{1 - (\xi_1\xi_2 + \eta_1\eta_2 + \zeta_1\zeta_2)}{1 + (\xi_1\xi_2 + \eta_1\eta_2 + \zeta_1\zeta_2)} = \frac{1 - \cos\omega_{12}}{1 + \cos\omega_{12}},$$

whence

(15.1) $$f_1(z_1, z_2) = \left|\tan \tfrac{1}{2}\tilde{D}(z_1, z_2)\right|.$$

Let us denote by $\tan^{-1}u$ the principal value of the function arc tan u; then by (15.1)

(15.11) $$\tilde{D}(z_1, z_2) = \pm 2 \tan^{-1} f_1(z_1, z_2) + 2k\pi.$$

As 'principal value' of the distance between z_1 and z_2 we are thus led to define

(15.12) $$D(z_1, z_2) = 2 \tan^{-1} f_1(z_1, z_2).$$

Then

(15.13) $$0 \leqslant D(z_1, z_2) = \omega_{12} < \pi,$$

$$D(z_1, z_2) = 0 \quad \text{if and only if} \quad z_1 = z_2,$$

and $D(z_1, z_2)$ will approach the value π if z_2 approaches the antipodal point $-1/\bar{z}_1$ of z_1.

b. *Additivity and triangle inequality.* We shall say that three points $\mathbf{P}_1, \mathbf{P}_2, \mathbf{P}_3$ on a great circle of the sphere lie 'in the order of the indices' if $\mathbf{P}_2$ and its diametrically opposite point $-\mathbf{P}_2$ separate $\mathbf{P}_1$ and $\mathbf{P}_3$. Correspondingly, three points z_1, z_2, z_3 on a spherical straight line are said to lie in the order of the indices if z_2 and its antipodal point $-1/\bar{z}_2$ separate z_1 and z_3 on the circle which represents the spherical straight line.

For geometrical reasons it is obvious that if z_1, z_2, z_3 lie in the order of the indices on their spherical straight line, then

$$\tilde{D}(z_1, z_2) + \tilde{D}(z_2, z_3) = \pm \tilde{D}(z_1, z_3) + 2k\pi$$

where k is an integer.

A stricter additivity is readily established for the principal value $D(z_1, z_2)$ of the distance. With regard to the situation on the sphere it is readily seen that if z_1, z_2, z_3 lie in the order of the indices on a spherical straight line then

$$(15.2)\qquad D(z_1, z_2)+D(z_2, z_3) = \begin{cases} D(z_1, z_3) & \text{if } D(z_1, z_2)+D(z_2, z_3) \leqslant \pi \\ 2\pi - D(z_1, z_3) & \text{if } D(z_1, z_2)+D(z_2, z_3) \geqslant \pi. \end{cases}$$

The triangle inequality is contained in the following

Theorem A. *For any three points z_1, z_2, z_3 in the spherical plane*

$$D(z_1, z_3) \leqslant D(z_1, z_2)+D(z_2, z_3) \leqslant 2\pi - D(z_1, z_3).$$

The sign of equality is valid if and only if z_1, z_2, z_3 lie on a spherical straight line in the order of their indices and

$$D(z_1, z_2)+D(z_2, z_3)\begin{cases} \leqslant \pi & \text{in the first case} \\ \geqslant \pi & \text{in the second case.} \end{cases}$$

Proof. There is a spherical motion (rotational Moebius transformation) $\mathfrak{H}$ such that

$$\mathfrak{H}(z_1) = x > 0, \qquad \mathfrak{H}(z_2) = 0, \qquad \mathfrak{H}(z_3) = z.$$

As the motion does not change the spherical distances, it will be sufficient to prove that

$$(15.3)\qquad D(x, z) \leqslant D(x, 0)+D(0, z) \leqslant 2\pi - D(x, z)$$

with the sign of equality if and only if z is real and negative and

$$1+xz\begin{cases} > 0 & \text{in the first case} \\ < 0 & \text{in the second case.} \end{cases}$$

Now $D(x, 0) = 2\tan^{-1}x$, $D(0, z) = 2\tan^{-1}|z|$ and

$$\tan^{-1} x+\tan^{-1}|z| = \begin{cases} \tan^{-1}\dfrac{x+|z|}{1-x|z|} & \text{if } 1-x|z| > 0 \\ \dfrac{\pi}{2} & \text{if } 1-x|z| = 0 \\ -\tan^{-1}\dfrac{x+|z|}{|1-x|z||}+\pi & \text{if } 1-x|z| < 0. \end{cases}$$

We examine the three cases separately.

(i) $x|z| < 1$. One has to verify that

$$(15.31)\qquad \left|\frac{x-z}{1+xz}\right| \leqslant \frac{x+|z|}{1-x|z|}$$

which is easy because $|x-z| \leqslant x+|z|$ and $|1+xz| > 1-x|z|$. Equality in (15.31), that is, $|x-z|(1-x|z|) = (x+|z|)|1+xz|$, is seen to be equivalent to the condition $2|z|+z+\bar{z} = 0$, that is, $\operatorname{Re} z = -|z|$. Hence z must lie on the negative real axis and $1+xz > 0$. The second inequality (15.3) is trivial.

(ii) $x|z| = 1$. In this case

$$\tan^{-1}\left|\frac{x-z}{1+xz}\right| \leqslant \frac{\pi}{2},$$

with equality if and only if $z = -1/x = x^*$.

(iii) $x|z| > 1$. Now the second inequality (15.3) is equivalent to

$$\tan^{-1}\frac{x+|z|}{|1-x|z||}+\tan^{-1}\left|\frac{x-z}{1+xz}\right| \leqslant \pi,$$

which is correct because each of the left-hand terms is less than or equal to $\pi/2$, both are equal to $\pi/2$ if and only if z is negative and $1+xz = 0$. In virtue of the assumption, the first inequality is trivial.

The discussion which at the end of § 14, **b**, led to the introduction of the hyperbolic length of the arc $\mathscr{C}$ of a curve in the hyperbolic plane can now be repeated *mutatis mutandis* leading to the spherical length of an arc of a curve in the spherical plane. By means of the triangle inequality it is then possible to establish the minimum length property of the spherical straight line. The proof is verbally the same as the one given for Theorem B in § 14, **b**. It will not be repeated here.

c. *Spherical circles.* Spherical circles, that is, $\mathscr{R}$-circles according to § 13, **d**, are ordinary real circles in the complex plane. Suppose the circle $\mathfrak{C}$ is generated by a rotational Moebius transformation $\mathfrak{H}$ as indicated by (13.1). The fixed points z_0, z_0^* of $\mathfrak{H}$ are then the spherical centres of $\mathfrak{C}$. These two points are antipodals and mutually inverse (symmetric) with respect to the circle $\mathfrak{C}$. All circles spherically concentric with the circle $\mathfrak{C}$ form the hyperbolic pencil of which z_0 and z_0^* are the point circles.

Let z be any point on the circle $\mathfrak{C}$ with the spherical centres z_0, z_0^*. The two non-negative numbers

$$r = D(z_0, z) \leqslant \pi, \qquad r^* = D(z_0^*, z) \leqslant \pi$$

which satisfy the condition

$$r+r^* = \pi,$$

are the spherical radii of $\mathfrak{C}$; the smaller one of the two may be called *the* spherical radius of $\mathfrak{C}$. It is always $\leqslant \pi/2$. The corresponding centre may be called *the* spherical centre of $\mathfrak{C}$.

The centre remains ambiguous if and only if $r = r^* = \pi/2$ in which case the spherical circle $\mathfrak{C}$ is a spherical straight line. Conversely a spherical straight line may be defined as a circle with the spherical radius $\pi/2$.

Spherical straight lines are euclidean straight lines if and only if they pass through the point 0. Spherically concentric circles are euclidean concentric if and only if they have 0 as centre. Among the concentric circles about 0 there is exactly one spherical straight line, namely the unit circle. Any two of its diametrically opposite points are antipodes. This is not so for any other spherical straight line. In every pencil of spherically concentric circles there is one and only one spherical straight line.

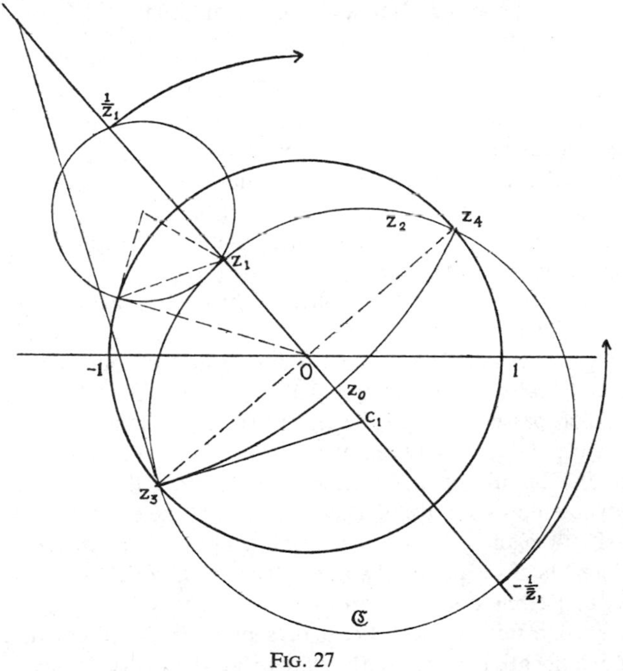

FIG. 27

The elliptic pencil of circles through the two (mutually antipodal) centres of $\mathfrak{C}$ is the pencil of spherical diameters of the circles of the orthogonal hyperbolic pencil of spherically concentric spherical circles. All these diameter circles are spherical straight lines. One of them passes through 0 and is therefore an ordinary straight line $\mathfrak{l}$ which also passes through the euclidean centre of $\mathfrak{C}$. On this line we can construct the spherical centre z_0 of $\mathfrak{C}$. Let z_1 be a point on $\mathfrak{C}$, but not on $\mathfrak{l}$. Construct the ortho-circle $\mathfrak{C}_1$ through z_1 and its antipodal $z_1^* = -1/\bar{z}_1$; it meets the line $\mathfrak{l}$ at the spherical centres z_0 and $z_0^* = -1/\bar{z}_0$ of $\mathfrak{C}$.

By a similar construction one finds the spherical straight line passing through two given points z_1, z_2. It is the circle $\mathfrak{C}$ through the three points $z_1, z_2, z_1^* = -1/\bar{z}_1$. The one spherical centre z_0 of $\mathfrak{C}$ is obtained as the intersection of two spherical diameters; the one through 0 (an ordinary straight line, cf. Figure 27), the other one through the antipodal points z_3, z_3^* where the circle $\mathfrak{C}$ meets the unit circle.

A discussion in spherical geometry is often simplified by transforming a figure into a 'normal position' by means of a certain spherical motion. Thus a circle can be transformed into a position where its centre coincides with 0. If ρ is its euclidean radius then its spherical radius is determined as

$$r = 2\,\text{Min}\left(\tan^{-1}\rho, \frac{\pi}{2} - \tan^{-1}\rho\right). \tag{15.4}$$

d. *Elliptic geometry.* 'It was Klein who first saw clearly how to rid spherical geometry of its one blemish: the fact that two coplanar lines (being two great circles of a sphere) have not just one, but two common points. Since every point determines a unique antipodal point, and every figure is thus duplicated at the antipodes, he realized that nothing would be lost, but much be gained by abstractly identifying each pair of antipodal points, i.e. by changing the meaning of the word "point" so as to call such a pair one point.'[14]

By this process of identification spherical geometry is changed into elliptic geometry. This identification of any two antipodal points z and $z^* = -1/\bar{z}$, or of the stereographically corresponding points **P** and **P*** on the sphere, may be realized geometrically by joining **P** and **P*** by the (ordinary) straight line **PP***, which of course passes through the centre **O** of the sphere. Thus we have a one–one correspondence between the points of the 'elliptic plane' and the elements of the bundle of all straight lines through the point **O**, every line through **O** being representative of a certain point of the elliptic plane. A straight line in this elliptic plane will then be represented by a plane through **O**, its points being the elements of the pencil of all lines through **O** and the pairs of antipodal points of a great circle on the sphere.

We may go one step further. Let π be a plane in space not through **O**, for instance the tangential plane to the sphere at the North pole, given by the equation $\zeta = 1$. Each line through **O**, except those in the equatorial plane $\zeta = 0$, meets this plane at a certain point, with the coordinates X, Y, 1, so that

$$X + iY = \frac{\xi + i\eta}{\zeta} = \frac{2z}{1 - z\bar{z}} \tag{15.5}$$

[cf. (3.3)].

Every plane through **O**, except the equatorial plane, cuts the plane π in a well-defined straight line. If we complete the plane π by a 'straight line at infinity' as carrier of the points corresponding to the lines through **O** in the equatorial plane, we obtain another model of the elliptic plane, namely the *real projective plane*.

[14] Quoted from Coxeter [1], p. 13.

The elements, points, and straight lines in this plane, are the same for both projective and elliptic geometry. Elliptic geometry is obtained within the framework of projective geometry by introducing a 'metric' or distance, viz. the spherical distance, as a two-point invariant of the group of motions. In the completed plane these are represented by Moebius transformations leaving invariant the imaginary unit circle $z\bar{z}+1 = 0$. Since by (15.5)

$$X^2+Y^2+1 = \left(\frac{1+z\bar{z}}{1-z\bar{z}}\right)^2,$$

the group of 'elliptic motions' in π is represented by the group of all projective transformations (collineations) leaving invariant again the imaginary unit circle $X^2+Y^2+1 = 0$, now, however, attached to the plane π.

Thus the elliptic metric in the projective plane π is defined by this invariant conic section or 'absolute', represented by the imaginary unit circle. In § 14, ex. 2, it has been pointed out that hyperbolic geometry can also be realized in the real projective plane if another conic section, namely the real unit circle, is taken as invariant or absolute. In hyperbolic geometry this absolute played the role of the infinite of the hyperbolic plane. In analogy it may be said that the infinite of the elliptic plane is imaginary.

The fact that there is no real infinity in elliptic geometry is further indicated by the boundedness (15.13) of the distance D, which indeed provides the proper metric in the 'conformal model' of the elliptic plane, obtained from the completed plane of the complex numbers by the identification of antipodes.

It may be pointed out that also in projective geometry the 'straight line at infinity' whose adjunction to the euclidean plane turns the latter into the usual primitive model of the projective plane, is not to be considered as a distinguished element in the projective plane. Projective geometry requires its 'naturalization' so that in the axioms it is not distinguished from any other line in the projective plane.

The elliptic distance of two points $(X_1, Y_1, 1)$, $(X_2, Y_2, 1)$ in the plane π is measured by the spherical distance $D(z_1, z_2)$ of the two corresponding points z_1, z_2 in the completed plane, or the angle ω_{12} between the two vectors $\overrightarrow{\mathbf{OP}}_1$, $\overrightarrow{\mathbf{OP}}_2$. It is therefore obtained from

$$\text{(15.6)} \qquad \cos\omega_{12} = \xi_1\xi_2+\eta_1\eta_2+\zeta_1\zeta_2 = \frac{X_1X_2+Y_1Y_2+1}{\sqrt{(X_1{}^2+Y_1{}^2+1)}\sqrt{(X_2{}^2+Y_2{}^2+1)}}.$$

For further information concerning the relations between the different branches of geometry and their realizations, reference must be made to the literature.[15]

[15] Cf. Coxeter [1], chapters I, VI, and XIV, where further references will be found. In particular Klein [2], chapter IV, § 5 may be mentioned. In addition Hilbert and Cohn-Vossen [1], § 34 inter alia is of interest here.

e. *Spherical trigonometry.* A spherical triangle is not at all uniquely defined by its three vertices z_1, z_2, z_3. Its sides are segments on the spherical straight lines through z_2, z_3; z_3, z_1; z_1, z_2. The triangle with the vertices z_1, z_2, z_3 will be uniquely defined if we introduce the restriction that it should be situated entirely on one hemisphere and that the lengths of its sides should coincide with

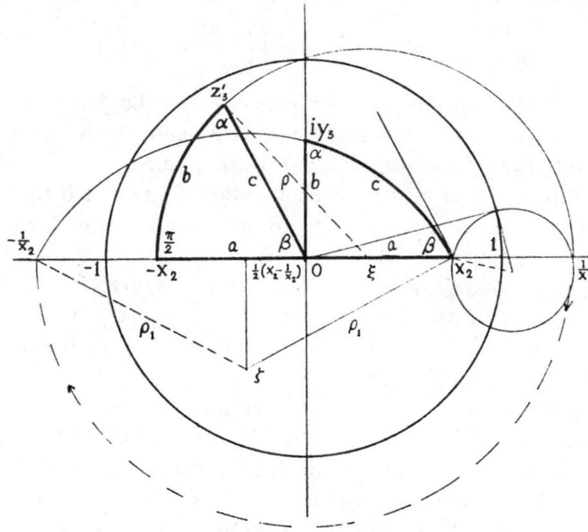

FIG. 28

the principal values of the distances $D(z_2, z_3)$, $D(z_3, z_1)$, $D(z_1, z_2)$ all to be less than π. Then the triangle inequalities

$$D(z_i, z_k) < D(z_i, z_j)+D(z_j, z_k)$$

are satisfied. Moreover we require that

$$D(z_2, z_3)+D(z_3, z_1)+D(z_1, z_2) < 2\pi.$$

Spherical triangles of this kind are called Euler triangles.

Under these conditions, complete formal analogy with the formulae of hyperbolic trigonometry (§ 14, **d**) will be found. The transition from hyperbolic to spherical trigonometry will be effected by replacing the hyperbolic functions in the relations of hyperbolic trigonometry by the corresponding trigonometric functions.

As in hyperbolic geometry, the whole of trigonometry depends on the theory of the right triangle. Taking over all notations and arrangements from § 14, **d**, we have the two normal positions of the right triangle as shown in Figure 28.

Referring to the first position (with the right angle at 0), we have, as in (14.5),

$$x_2 = \tan\frac{a}{2}, \qquad y_3 = \tan\frac{b}{2}, \qquad \frac{|x_2 - iy_3|}{|1 - ix_2y_3|} = \tan\frac{c}{2}, \tag{15.72}$$

whence

$$\tan^2\frac{c}{2} = \frac{\tan^2(a/2)+\tan^2(b/2)}{1+\tan^2(a/2)\tan^2(b/2)}$$

and, by making use of elementary trigonometrical formulae,

$$\cos c = \cos a \cos b. \tag{15.73}$$

This is the 'spherical Pythagoras' [cf. (14.52)] which indicates [as in the Remark under (14.52)] that spherical geometry is also 'locally euclidean.'

To obtain the relations between sides and angles of the right triangle, we move it into the second normal position with the angle β at the origin. The construction becomes evident from Figure 28. As a spherical segment the side b must now lie on the circle through the point $-x_2$ and its antipode $1/x_2$; the centre of this circle is the point

$$\xi = \tfrac{1}{2}\left(\frac{1}{x_2} - x_2\right) \tag{15.74}$$

on the real axis, and

$$\rho = \tfrac{1}{2}\left(\frac{1}{x_2} + x_2\right)$$

is its radius, so that $\rho^2 - \xi^2 = 1$. The circle then is defined by the equation $(x-\xi)^2 + y^2 = \rho^2$, or

$$z\bar{z} - 2\xi \operatorname{Re} z = 1.$$

On this circle lies the vertex

$$z_3' = r_3 e^{i\beta'} \qquad (\beta' = \pi - \beta)$$

of the triangle; hence $r_3{}^2 - 2\xi r_3 \cos\beta' = 1$, and thus by (15.74),

$$\frac{r_3}{1-r_3{}^2}\cos\beta = \frac{x_2}{1-x_2{}^2}.$$

Further,

$$x_2 = \tan\frac{a}{2}, \qquad r_3 = \tan\frac{c}{2}; \tag{15.75}$$

therefore

$$\tan c \cos\beta = \tan a, \qquad \tan c \cos\alpha = \tan b \tag{15.76}$$

[cf. (14.57)].

The spherical analogues of the relations (14.56) may now be obtained by a simple trigonometrical calculation using (15.73) twice:

$$\sin c \sin\beta = \sin c\left(1-\frac{\sin^2 a}{\cos^2 a}\frac{\cos^2 c}{\sin^2 c}\right)^{\frac{1}{2}} = \sqrt{(\sin^2 c - \sin^2 a \cos^2 b)},$$

whence

(15.77) $$\sin c \sin\beta = \sin b, \qquad \sin c \sin\alpha = \sin a.$$

We shall not write down the formulae for the general Euler triangle; it is now obvious how they can be obtained from the corresponding hyperbolic relations (14.58) and (14.59) (cf. Figure 29). We conclude with

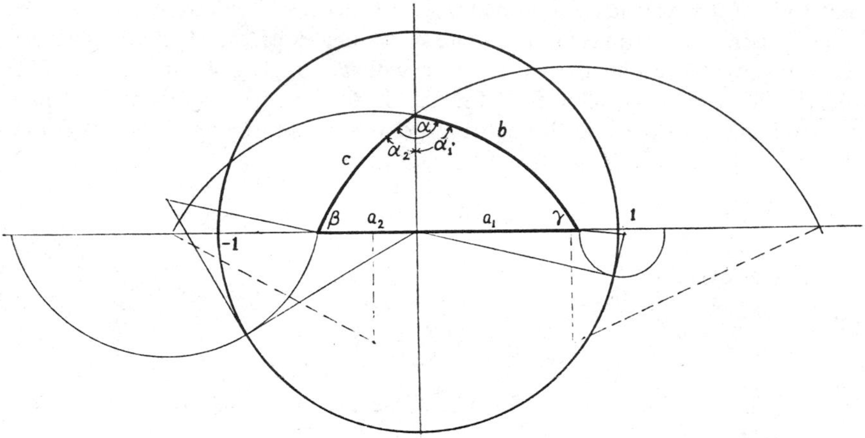

FIG. 29

Theorem E. *The sum of the angles*

$$\sigma = \alpha+\beta+\gamma$$

of an Euler triangle satisfies the inequalities

$$\pi < \sigma < 3\pi.$$

Proof. For a right triangle ($\gamma = \pi/2$),

$$\cos(\alpha+\beta) = \frac{\tan b}{\tan c}\frac{\tan a}{\tan c} - \frac{\sin a}{\sin c}\frac{\sin b}{\sin c}$$

$$= \frac{\sin b \sin a}{\sin^2 c}(\cos c - 1) < 0.$$

Indeed none of the sides of the triangle is $\geqslant \pi$, all the sine terms are therefore positive. Thus

$$\frac{\pi}{2} < \alpha+\beta < 3\frac{\pi}{2}. \tag{15.78}$$

As for the general Euler triangle, we may divide it into two right triangles, for instance by the height perpendicular to the side a (cf. Figure 29). This height divides the angle α into α_1 and α_2, and the theorem follows by adding the relations (15.78) for the two right triangles:

$$\frac{\pi}{2} < \alpha_1+\gamma < 3\frac{\pi}{2}, \qquad \frac{\pi}{2} < \alpha_2+\beta < 3\frac{\pi}{2}.$$

The (always positive) magnitude

$$\varepsilon = \alpha+\beta+\gamma-\pi \tag{15.79}$$

is called the 'spherical excess' of an Euler triangle. Like the defect in the case of a hyperbolic triangle, the excess represents a measure of the area of the triangle (cf. ex. 9).

Examples

1. An inversion with respect to a spherical straight line will be called a spherical symmetry. A spherical symmetry leaves invariant the imaginary unit circle; it maps spherical straight lines onto spherical straight lines, and it leaves invariant the spherical distance $D(z_1, z_2)$ of two points (cf. § 14, ex. 3).

2. For a given spherical segment (z_1, z_2), find by geometrical construction the spherically congruent segment in 'normal position,' that is, so that 0 is one endpoint while the other endpoint lies on the positive real axis. Likewise for a given triangle find the spherically congruent triangle in a normal position (cf. § 14, ex. 4).

3. Prove the uniqueness, up to a positive factor, of the spherical distance D, if it is required to be

1. Invariant by spherical motion.
2. Non-negative.
3. Additive on every spherical straight line on one hemisphere (cf. Theorem C).

As in § 14, ex. 5, the problem may be reduced to the discussion of the functional equation

$$\phi\left(\frac{x_1-x_2}{1+x_1x_2}\right) = \phi(x_1)-\phi(x_2)$$

for the function $\phi(u)$ such that $D(z_1, z_2) = \phi[f_1(z_1, z_2)]$.

4. *Distance in the projective model of elliptic geometry* (cf. § 15, **d**). In the projective plane π, touching the sphere at the north pole, let **Q** be the image of

the point $\mathbf{P}(\xi, \eta, \zeta)$ on the sphere. We may take ξ, η, ζ as normalized homogeneous coordinates of $\mathbf{Q}$, normalization consisting in the condition $\xi^2+\eta^2+\zeta^2 = 1$. It is known from vector algebra that the scalar product

$$(\xi_1\xi_2) = \xi_1\xi_2+\eta_1\eta_2+\zeta_1\zeta_2$$

is the only independent two-point invariant with respect to rotations of the sphere about $\mathbf{O}$. Thus the elliptic distance of two points $\mathbf{Q}_1$, $\mathbf{Q}_2$ in π can be represented in the form

$$D(\mathbf{Q}_1, \mathbf{Q}_2) = \phi[(\xi_1\xi_2)],$$

where $\phi(u)$ is a non-negative continuous monotonic function of u. By (15.6)

$$(\xi_1\xi_2)^2 \leqslant 1,$$

which is also a consequence of the Cauchy-Schwarz inequality:

$$(\xi_1\xi_2)^2 \leqslant (\xi_1\xi_1)(\xi_2\xi_2).$$

The additivity of D implies the relation

$$\phi[(\xi_1\xi_2)]+\phi[(\xi_2\xi_3)] = \phi[(\xi_1\xi_3)] \tag{15.8}$$

for three points $\mathbf{Q}_1, \mathbf{Q}_2, \mathbf{Q}_3$ on a spherical straight line in this order. By invariance we may assume this line to be the real axis, and

$$\begin{aligned} \mathbf{Q}_2 &: \xi_2 = 0, & \eta_2 &= 0, & \zeta_2 &= 1 \\ \mathbf{Q}_1 &: \xi_1 < 0, & \eta_1 &= 0, & \zeta_1 &= \sqrt{(1-\xi_1^2)} \\ \mathbf{Q}_3 &: \xi_3 > 0, & \eta_3 &= 0, & \zeta_3 &= \sqrt{(1-\xi_3^2)}. \end{aligned}$$

Thus by (15.8),

$$\phi[\sqrt{(1-\xi_1^2)}]+\phi[\sqrt{(1-\xi_3^2)}] = \phi[\xi_1\xi_3+\sqrt{(1-\xi_1^2)}\sqrt{(1-\xi_3^2)}]$$

or, if

$$\begin{aligned} 1-\xi_1^2 &= u^2, & \xi_1 &= -\sqrt{(1-u^2)} \\ 1-\xi_3^2 &= v^2; & \xi_3 &= \sqrt{(1-v^2)}, \end{aligned}$$

then

$$\phi(u)+\phi(v) = \phi[uv-\sqrt{(1-u^2)}\sqrt{(1-v^2)}].$$

The only continuous solution $\phi(u)$ of this functional equation which is positive and monotonic decreasing over the interval $-1 < u < 1$ is known to be c arc cos u, where c is a positive constant.

5. It might be of interest to sketch the treatment of the corresponding question in the case of the *projective model of hyperbolic geometry* which has been introduced in § 14, ex. 2. The hyperbolic plane, (that is, the interior of the unit circle in which the straight line segments represent the hyperbolic lines) may be placed into the projective plane π touching the sphere at $\mathbf{N}$ so that it appears

as the image of the northern hemisphere by projection parallel to the ζ-axis. A point **Q** in the hyperbolic plane with the homogeneous coordinates X, Y, Z will then be determined by

$$\frac{X}{Z} = \xi, \quad \frac{Y}{Z} = \eta \qquad (\xi^2+\eta^2+\zeta^2 = 1).$$

The equation of the unit circle (that is, the horizon, or absolute, or infinite, of the hyperbolic plane) is

$$X^2+Y^2-Z^2 = 0.$$

All its inner points are characterized by the inequality

$$X^2+Y^2-Z^2 < 0.$$

Thus we may normalize the homogeneous coordinates by putting

$$X^2+Y^2-Z^2 = -1. \tag{15.81}$$

The hyperbolic motions in π are represented by three-dimensional Lorentz transformations (cf. § 12, ex. 4) whose only independent two-point invariant is given by the 'Lorentz-scalar product'

$$[X_1X_2] = X_1X_2+Y_1Y_2-Z_1Z_2.$$

The hyperbolic distance of two points $\mathbf{Q}_1$, $\mathbf{Q}_2$ will appear in the form

$$D(\mathbf{Q}_1, \mathbf{Q}_2) = \phi([X_1X_2]),$$

where $\phi(u)$ is a non-negative continuous monotonic function of u.

Now we observe that

$$[X_1X_2]^2 \geqslant 1, \tag{15.82}$$

which is a consequence of the following 'Lorentz analogue' of the Cauchy-Schwarz inequality. Without regard to (15.81),

$$[X_1X_2]^2 \geqslant [X_1X_1][X_2X_2] \quad \text{if} \quad [X_1X_1] < 0,$$

with the sign of equality if and only if X_1, Y_1, Z_1 and X_2, Y_2, Z_2 are proportional. To prove this inequality we consider the following quadratic function in the real variable λ:

$$\begin{aligned}[(\lambda X_1+X_2)(\lambda X_1+X_2)] &= (\lambda X_1+X_2)^2+(\lambda Y_1+Y_2)^2-(\lambda Z_1+Z_2)^2 \\ &= [X_1X_1]\lambda^2+2[X_1X_2]\lambda+[X_2X_2].\end{aligned}$$

This is negative for large λ, but positive or zero for $\lambda = -Z_2/Z_1$. (Notice that $Z_1 \neq 0$ because $X_1^2+Y_1^2-Z_1^2 < 0$.) In the latter case the function has indeed the value

$$\left(X_2-\frac{Z_2}{Z_1}X_1\right)^2+\left(Y_2-\frac{Z_2}{Z_1}Y_1\right)^2 \geqslant 0.$$

Thus the quadratic function in λ has the discriminant

$$[X_1X_2]^2-[X_1X_1][X_2X_2] \geqslant 0,$$

and is equal to zero if and only if $X_2/X_1 = Y_2/Y_1 = Z_2/Z_1$. (Aczél [1].)

The additivity of the hyperbolic distance D implies

$$\phi(-[X_1X_2])+\phi(-[X_2X_3]) = \phi(-[X_1X_3]) \tag{15.83}$$

for any three points $\mathbf{Q}_1$, $\mathbf{Q}_2$, $\mathbf{Q}_3$ in this order on a hyperbolic straight line. By invariance we may assume this line to be the segment $-1 < X < 1$, and thus

$$\begin{aligned}
\mathbf{Q}_2 &: X_2 = 0, \quad Y_2 = 0, \quad Z_2 = 1\\
\mathbf{Q}_1 &: X_1 < 0, \quad Y_1 = 0, \quad Z_1 = \sqrt{(1+X_1^2)}\\
\mathbf{Q}_3 &: X_3 > 0, \quad Y_3 = 0, \quad Z_3 = \sqrt{(1+X_3^2)}.
\end{aligned}$$

Hence by (15.83)

$$\phi[\sqrt{(1+X_1^2)}]+\phi[\sqrt{(1+X_3^2)}] = \phi[-X_1X_3+\sqrt{(1+X_1^2)}\sqrt{(1+X_3^2)}],$$

or, if

$$\begin{aligned}
1+X_1^2 &= u, \quad X_1 = -\sqrt{(u^2-1)},\\
1+X_3^2 &= v, \quad X_3 = \sqrt{(v^2-1)},
\end{aligned}$$

then

$$\phi(u)+\phi(v) = \phi[uv+\sqrt{(u^2-1)}\sqrt{(v^2-1)}],$$

which for arguments $u > 1$ (cf. (15.82)) has as a continuous monotonic solution only the function $\phi(u) = c \operatorname{arch} u$, where c is a positive constant.

6. The locus of all points having constant spherical distance d from a spherical line $\mathfrak{l}$ (measured along spherical straight lines orthogonal to $\mathfrak{l}$) is a pair of circles $\mathfrak{C}$. Find the centre and radius of these circles as functions of d.

7. Show that three (positive) angles α, β, γ, all $< \pi$, which satisfy the inequalities of Theorem E, always define uniquely an Euler triangle up to a spherical motion or symmetry. This means that every Euler triangle with prescribed angles is congruent or symmetric to every other triangle with the same angles (cf. § 14, **f**, Theorem D).

8. Carrying through *mutatis mutandis* the discussion of § 14, ex. 11, in spherical geometry, the spherical area of a region $\mathscr{A}$ in the completed plane is found to be equal to

$$m(\mathscr{A}) = 4\iint_{\mathscr{A}} \frac{dx\,dy}{(1+|z|^2)^2}.$$

If $\mathscr{A}$ is a circular disc with the spherical radius r, we obtain its spherical area, $m(\mathscr{A}) = 4\pi \sin^2 \tfrac{1}{2}r$.

9. Prove that the spherical area of an Euler triangle is equal to its spherical excess. Verify that this magnitude has the typical properties of an area: additivity, and invariance with respect to spherical motions and symmetries.

10. The spherical length of a rectifiable curve $\mathscr{C}$ is given by the line integral

$$L = \int_{\mathscr{C}} \frac{2|dz|}{1+|z|^2} \qquad \text{(cf. § 14, ex. 13).}$$

Hence the spherical length of the circumference of a spherical circle $\mathfrak{C}$ of spherical radius r is equal to

$$L = 2\pi \sin r.$$

APPENDIX 1

UNIQUENESS OF THE CROSS RATIO

According to § 6, **d**, Theorem C (p. 47), the cross ratio of four points of the completed plane is an invariant of the group $\mathscr{M}$ of all Moebius transformations. It will now be shown that as such the cross ratio is essentially unique, that is

Theorem. *Every four-point invariant of the Moebius group $\mathscr{M}$ is a function of the cross ratio.* (*Cf. Aczél* [2], *Schwerdtfeger* [5].)

Proof. Let z_1, z_2, z_3 be three distinct constant points, e.g.,

$$z_1 = 0, \qquad z_2 = 1, \qquad z_3 = -1.$$

Their cross ratio with an arbitrary z is found to be

$$\lambda = (z, z_1; z_2, z_3) = -\frac{z-1}{z+1},$$

so that (6.35) appears in the form

$$(Z, Z_1; Z_2, Z_3) = -\frac{z-1}{z+1}.$$

By solving this equation with respect to Z we obtain $Z = \mathfrak{H}(z)$, the Moebius transformation which satisfies the conditions of Theorem D (p. 47), namely $\mathfrak{H}(0) = Z_1, \mathfrak{H}(1) = Z_2, \mathfrak{H}(-1) = Z_3$. We consider the inverse $z = \mathfrak{H}^{-1}(Z)$ which according to (6.36) is a function of the cross ratio λ; since $-(z - 1)/(z + 1)$ is involutory

$$z = -\frac{\lambda - 1}{\lambda + 1},$$

and as a function of λ a four-point invariant.

Suppose now that $f(Z, Z_1; Z_2, Z_3)$ denotes an arbitrary four-point invariant. Invariance of this function with respect to $\mathfrak{H}^{-1}$ implies that

$$\begin{aligned} f(Z, Z_1, Z_2, Z_3) &= f(\mathfrak{H}^{-1}(Z), \mathfrak{H}^{-1}(Z_1); \mathfrak{H}^{-1}(Z_2), \mathfrak{H}^{-1}(Z_3)) \\ &= f(z, 0, 1, -1) \\ &= f\left(-\frac{\lambda - 1}{\lambda + 1}, 0, 1, -1\right) \end{aligned}$$

which is in fact a function of λ.

Appendix 2

A THEOREM OF H. HARUKI

In a recent paper [2], Haruki has established an interesting property of orthogonal pencils of circles. Consider two mutually orthogonal parabolic pencils (cf. Figure 17, p. 64) or two mutually orthogonal pencils, one of which is hyperbolic and the other one elliptic (cf. Figure 18, p. 67). Each couple of circles of the one pencil determines, with each couple of circles of the orthogonal pencil, in the parabolic case two, and in the hyperbolic-elliptic case four circular quadrangles.

Theorem. *The four vertices of each of these quadrangles are cocyclic, that is, situated on one and the same circle.*

In order to prove this fact we subject the pencils either to an inversion or to a Moebius transformation which turns them onto their normal forms (cf. § 4, ex. 2). Thus in the parabolic case, shifting the common point of all the circles into the point at infinity, we obtain two pencils of mutually orthogonal parallel straight lines. The quadrangles are now rectangles whose vertices are obviously cocyclic; therefore the vertices of the circular quadrangles in the original pencils are also cocyclic (cf. § 2, Theorem B or § 6, Theorem B).

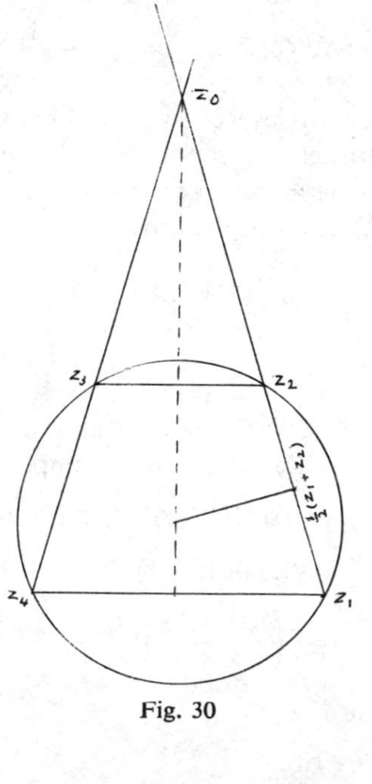

Fig. 30

In the hyperbolic-elliptic case the normal form consists of the elliptic pencil of straight lines through a point z_0 and the hyperbolic pencil of concentric circles about z_0. The vertices z_1, z_2, z_3, z_4 of a quadrangle are also the vertices of a symmetric trapezoid and therefore cocyclic (cf. Figure 30). Again the theorem is verified by returning from the normal form to the originally given pair of mutually orthogonal pencils.

It is readily seen that the theorem fails to be true if the two pencils in question are not orthogonal. Thus orthogonal pencils are characterized by the cocyclicity of their quadrangles.

APPENDIX 3

APPLICATIONS OF THE CHARACTERISTIC POLYNOMIAL

This section of the appendix is a sequel to § 8, **e**, p. 69. It deals with some results of the (unpublished) thesis of R. Starrost [1]. The author is indebted to Professor F. Bachmann for sending him a carbon copy of this paper which systematically uses the characteristic parallelogram as the basis for geometric constructions connected with the corresponding Moebius transformation.

a. *Construction of the image point.* Let $\mathfrak{H}$ be a non-integral Moebius transformation given by the four vertices z_∞, Z_∞, γ_1, γ_2 of its characteristic parallelogram which satisfy the condition

$$z_\infty + Z_\infty = \gamma_1 + \gamma_2.$$

According to (8.75)

$$Z = \mathfrak{H}(z) = \frac{Z_\infty z - \gamma_1\gamma_2}{z - z_\infty}. \tag{8.751}$$

This equation is equivalent to each of the following two relations:

$$Z - Z_\infty = \frac{(\gamma_1 - Z_\infty)(\gamma_1 - z_\infty)}{z - z}, \qquad Z - \gamma_2 = \frac{(\gamma_1 - z_\infty)(z - \gamma_2)}{z - z_\infty}$$

whence

$$\frac{\gamma_1 - z_\infty}{z - z_\infty} = \frac{Z - Z_\infty}{\gamma_1 - Z_\infty} = \frac{Z - \gamma_2}{z - \gamma_2}. \tag{8.76}$$

These relations, implicit representations of the Moebius transformation $\mathfrak{H}$, express the similarity of the three triangles

$$\Delta\, \gamma_1 z_\infty z, \qquad \Delta\, Z Z_\infty \gamma_1, \qquad \Delta\, Z \gamma_2 z.$$

These similarities enable us to construct graphically the image point $Z = \mathfrak{H}(z)$ for every given $z \neq z_\infty$ if the characteristic parallelogram of $\mathfrak{H}$ is known, cf. Figure 31. Indeed the first of the triangles is given. The figure indicates how the second triangle is obtained. The similarity of the third triangle may serve to check the accuracy of the construction.

b. Let $\mathfrak{H} = \mathfrak{J}$ where $\mathfrak{J}$ is an *involution*, that is $Z_\infty = z_\infty$. The characteristic parallelogram then degenerates into a segment, denoted by $\mathfrak{s} = \mathfrak{s}(\gamma_1, \gamma_2)$ whose

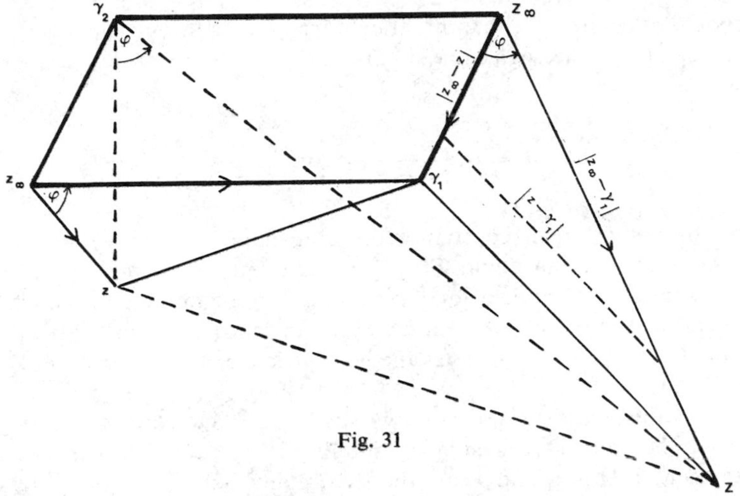

Fig. 31

endpoints are the fixed points γ_1, γ_2 of $\mathfrak{J}$ and the pole z_∞ is the midpoint of $\mathfrak{s}$ (cf. § 6, **f**, p. 50). The relations (8.76) now are

$$\frac{\gamma_1 - z_\infty}{z - z_\infty} = \frac{Z - z_\infty}{\gamma_1 - z_\infty} = \frac{Z - \gamma_2}{z - \gamma_2}. \tag{8.77}$$

Thus the triangles $\Delta\, \gamma_1 z_\infty z$, $\Delta\, Z z_\infty \gamma_1$, $\Delta\, Z \gamma_2 z$ are similar whence, as before, one can construct the image point $Z = \mathfrak{J}(z)$ for every given $z \neq z_\infty$, as shown in Figure 32.

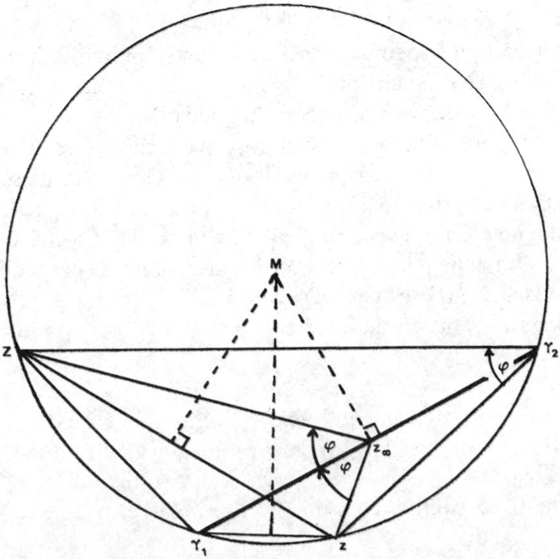

Fig. 32

It should be pointed out that the straight line passing through γ_1 and γ_2, that is the carrier line of the segment $\mathfrak{s}$, is the bisectrice of the angle $\sphericalangle Zz_\infty z = 2\varphi$ where

$$\varphi = \text{arc}\,\frac{Z - z_\infty}{\gamma_1 - z_\infty} = \text{arc}\,\frac{Z - \gamma_2}{z - \gamma_2}.$$

Since the characteristic constant k of an involution equals -1 we conclude from (8.42) that the four points Z, z, γ_1, γ_2 have real cross ratio; therefore they lie on a common real circle or on a straight line, the latter if, and only if, z lies on the carrier line of the segment $\mathfrak{s}(\gamma_1, \gamma_2)$. In this case γ_1 (or γ_2) divides the segment $\mathfrak{s}(z, Z)$ so that $\gamma_1 - z_\infty$ is the geometric mean of $z - z_\infty$ and $Z - z_\infty$ (cf. (8.77)).

Inverting the preceding discussion we shall now show that for every segment $\mathfrak{s}(\gamma_1, \gamma_2)$ in the plane there is an involution $\mathfrak{J}$ which has $\mathfrak{s}(\gamma_1, \gamma_2)$ as its characteristic segment, that is γ_1 and γ_2 are the fixed points and $z_\infty = \frac{1}{2}(\gamma_1 + \gamma_2)$ is the pole of $\mathfrak{J}$. Indeed if these three points are given $\mathfrak{J}$ is explicitly given by

$$Z = \mathfrak{J}(z) = \frac{z_\infty z - \gamma_1\gamma_2}{z - z_\infty}.$$

It is also the Moebius transformation defined by the three conditions

$$\mathfrak{J}(\gamma_1) = \gamma_1, \qquad \mathfrak{J}(\gamma_2) = \gamma_2, \qquad \mathfrak{J}(z_\infty) = \infty \qquad \text{(cf. § 6, \textbf{d})}.$$

By § 6, Theorem F, a Moebius transformation is an involution if it has a pair of conjugate points z, Z, such that $Z = \mathfrak{J}(z)$ and $z = \mathfrak{J}(Z)$. Moreover all the pairs $z, \mathfrak{J}(z)$ are conjugate with respect to $\mathfrak{J}$. One other pair of conjugates defines an involution; in particular we may choose the pair z_∞, ∞, that is: *A pair of conjugates z, Z and the pole z_∞ define an involution* $\mathfrak{J}$ and also its segment $\mathfrak{s}(\gamma_1\gamma_2)$.

Geometrically: Referring to Figure 32 let the points z, Z, z_∞ be given. The line through the now unknown points γ_1, γ_2 will be the bisectrice $\mathfrak{b}$ of the angle $\sphericalangle zz_\infty Z = 2\varphi$. Draw the perpendicular on $\mathfrak{b}$ at z_∞ and the perpendicular median to the segment $\mathfrak{s}(z, Z)$; these two lines will meet at a point M. Draw the circle about M through z. This circle intersects $\mathfrak{b}$ at the points γ_1 and γ_2.

c. *Decomposition of a Moebius transformation.* Let $\mathfrak{H}$ be a non-integral Moebius transformation, neither involutory nor parabolic. From § 8, ex. 9, we conclude that $\mathfrak{H}$ can be written as the product of two involutions $\mathfrak{J}', \mathfrak{J}''$. Indeed, let γ_1, γ_2 be the fixed points, z_∞ and Z_∞ the poles of $\mathfrak{H}$ and $\mathfrak{H}^{-1}$. Select an involution $\mathfrak{J}'$ for which γ_1, γ_2 are a pair of conjugate points: $\mathfrak{J}'(\gamma_1) = \gamma_2$. Then $\mathfrak{J}'' = \mathfrak{H}\mathfrak{J}'$ is an involution. Hence

$$\mathfrak{H} = \mathfrak{J}''\mathfrak{J}'$$

is a 'decomposition' of $\mathfrak{H}$.

Given the characteristic parallelogram of $\mathfrak{H}$ we wish to determine the characteristic segments $\mathfrak{s}' = \mathfrak{s}(\gamma_1', \gamma_2')$ and $\mathfrak{s}'' = \mathfrak{s}(\gamma_1'', \gamma_2'')$, and thereby the poles $z' = \frac{1}{2}(\gamma_1' + \gamma_2')$ and $z'' = \frac{1}{2}(\gamma_1'' + \gamma_2'')$ of $\mathfrak{J}'$ and $\mathfrak{J}''$, respectively. We call $\mathfrak{s}'$, $\mathfrak{s}''$ a decomposition of the parallelogram of $\mathfrak{H}$.

The pole z'_∞ of $\mathfrak{J}'$ may be chosen ad libitum, different from each of the four given points $\gamma_1, \gamma_2, z_\infty, Z_\infty$, the fixed points and poles of $\mathfrak{H}$. So let z'_∞ coincide with the centre of the parallelogram of $\mathfrak{H}$:

$$z'_\infty = \tfrac{1}{2}(\gamma_1 + \gamma_2) = \tfrac{1}{2}(z_\infty + Z_\infty) \qquad \text{(cf. Figure 33).}$$

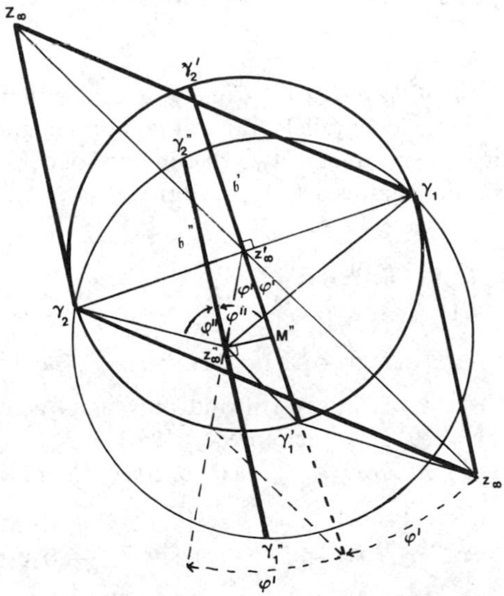

Fig. 33

The bisectrice $\mathfrak{b}'$ (corresponding to $\mathfrak{b}$ in the construction described in the last paragraph of subsection **b)** is the perpendicular median to $\mathfrak{s}(\gamma_1, \gamma_2)$ at z'_∞ since γ_1, γ_2 are conjugate with respect to $\mathfrak{J}'$ and $\sphericalangle \gamma_1 z'_\infty \gamma_2 = \pi$. The centre M' will be found as the intersection of the perpendicular median to $\mathfrak{s}(\gamma_1, \gamma_2)$ and the perpendicular of $\mathfrak{b}'$ at z'_∞; thus $M' = z'_\infty$. The circle about z'_∞ passing through γ_1 and γ_2 will meet $\mathfrak{b}'$ at the points γ_1', γ_2', the two fixed points of $\mathfrak{J}'$.

Now we observe that

$$\mathfrak{H}(z_\infty) = \mathfrak{J}''\mathfrak{J}'(z_\infty) = \infty$$

hence

$$\mathfrak{Z}'(z_\infty) = \mathfrak{Z}''(\infty) = z''_\infty.$$

By the construction described first in connection with Figure 32 the pole z''_∞ of $\mathfrak{Z}''$ is therefore found as the image of z_∞ under the mapping $\mathfrak{Z}'$. Thereby

$$\sphericalangle z_\infty z'_\infty \gamma'_1 = \sphericalangle \gamma'_1 z'_\infty z''_\infty = \varphi'$$

and the triangles $\Delta\ z_\infty z'_\infty \gamma'_1$ and $\Delta\ \gamma'_1 z'_\infty z''_\infty$ are similar. Further we note that $\mathfrak{Z}''\mathfrak{Z}'(\gamma_1) = \gamma_1$; therefore $\mathfrak{Z}''(\gamma_1) = \mathfrak{Z}'(\gamma_1) = \gamma_2$ so that γ_1 and γ_2 are conjugate also with respect to the involution $\mathfrak{Z}''$ of which we so far know only the pole z''_∞. To determine the fixed points γ''_1, γ''_2 we proceed as prescribed in Figure 33: Bisect the angle $\sphericalangle \gamma_1 z''_\infty \gamma_2 = 2\varphi''$ by the line $\mathfrak{b}''$, draw the perpendicular $\mathfrak{p}''$ to $\mathfrak{b}''$ at z''_∞ and determine the intersection M'' of $\mathfrak{p}''$ with $\mathfrak{b}'$, i.e., the perpendicular median to $\mathfrak{s}(\gamma_1, \gamma_2)$ at z'_∞. The circle about M'' through γ_1 intersects $\mathfrak{b}''$ at the points γ''_1, γ''_2.

d. *The parellelogram of the product.* Finally we deal with the problem of constructing the characteristic parallelogram of a non-integral Moebius transformation $\mathfrak{H}$, which is the product of two non-integral Moebius transformations $\mathfrak{H}_1$ and $\mathfrak{H}_2$, whose parallelograms are given, so that $\mathfrak{H} = \mathfrak{H}_2\mathfrak{H}_1$. With reference to (8.75) we write

$$\mathfrak{H}_1 = \begin{pmatrix} Z_{1\infty} & -\gamma_{11}\gamma_{12} \\ 1 & -z_{1\infty} \end{pmatrix}, \qquad \mathfrak{H}_2 = \begin{pmatrix} Z_{2\infty} & -\gamma_{21}\gamma_{22} \\ 1 & -z_{2\infty} \end{pmatrix}$$

$$\gamma_{11} + \gamma_{12} = z_{1\infty} + Z_{1\infty}, \qquad \gamma_{21} + \gamma_{22} = z_{2\infty} + Z_{2\infty}.$$

The mapping $\mathfrak{H} = \mathfrak{H}_2\mathfrak{H}_1$ is integral if and only if its pole $z_\infty = \infty$, that is $\mathfrak{H}_2\mathfrak{H}_1(\infty) = \mathfrak{H}_2(Z_{1\infty}) = \infty$ and therefore $\mathfrak{H}_2^{-1}(\infty) = z_{2\infty} = Z_{1\infty}$. Thus $\mathfrak{H}$ is non-integral if and only if $Z_{1\infty} \neq z_{2\infty}$. To construct the poles z_∞ and Z_∞ of $\mathfrak{H}$ we note that

$$z_\infty = \mathfrak{H}^{-1}(\infty) = \mathfrak{H}_1^{-1}\mathfrak{H}_2^{-1}(\infty) = \mathfrak{H}_1^{-1}(z_{2\infty}), \qquad Z_\infty = \mathfrak{H}_2\mathfrak{H}_1(\infty) = \mathfrak{H}_2(Z_{1\infty})$$

and we use the method described in subsection **a** in connection with Fig. 31. The pole z_∞ of $\mathfrak{H}$ is the image of the (known) pole $z_{2\infty}$ of $\mathfrak{H}_2$ by the mapping $\mathfrak{H}_1^{-1}$ whose characteristic parallelogram is the parallelogram of $\mathfrak{H}_1$ with the interpretations of $z_{1\infty}$ and $Z_{1\infty}$ interchanged. By the same method we get Z_∞ as the image of $Z_{1\infty}$ by the mapping $\mathfrak{H}_2$.

Having thus constructed the two pole vertices of the parallelogram of $\mathfrak{H}$ we now introduce the midpoint of the segment $\mathfrak{s}(z_\infty, Z_\infty)$, namely

$$z^{(m)} = \tfrac{1}{2}(z_\infty + Z_\infty) = \tfrac{1}{2}(\gamma_1 + \gamma_2)$$

which will be important for the construction of the fixed points γ_1 and γ_2 of $\mathfrak{H}$.

By applying the mapping $\mathfrak{H}_1$ to $z^{(m)}$ and $\mathfrak{H}_2$ to $\mathfrak{H}_1(z^{(m)})$ we obtain (again graphically) the point

$$Z^{(m)} = \mathfrak{H}(z^{(m)}).$$

For the construction of γ_1 and γ_2 we recall that they form a pair of conjugate points with respect to each of the involution factors of a decomposition of $\mathfrak{H}$. We choose as $\mathfrak{J}'$ the involution which has $z^{(m)}$ as its pole: $z^{(m)} = z'_\infty$, and for which $z_\infty, Z^{(m)}$ is a pair of conjugate points. The characteristic segment $\mathfrak{s}(\gamma'_1, \gamma'_2)$ of $\mathfrak{J}'$ lies on the bisectrice $\mathfrak{b}'$ of the angle $\sphericalangle z_\infty z^{(m)} Z^{(m)} = 2\varphi'$ (cf. Figure 34). The fixed points γ'_1, γ'_2 of $\mathfrak{J}'$ are then found as follows: Draw the perpendicular $\mathfrak{p}'$ to

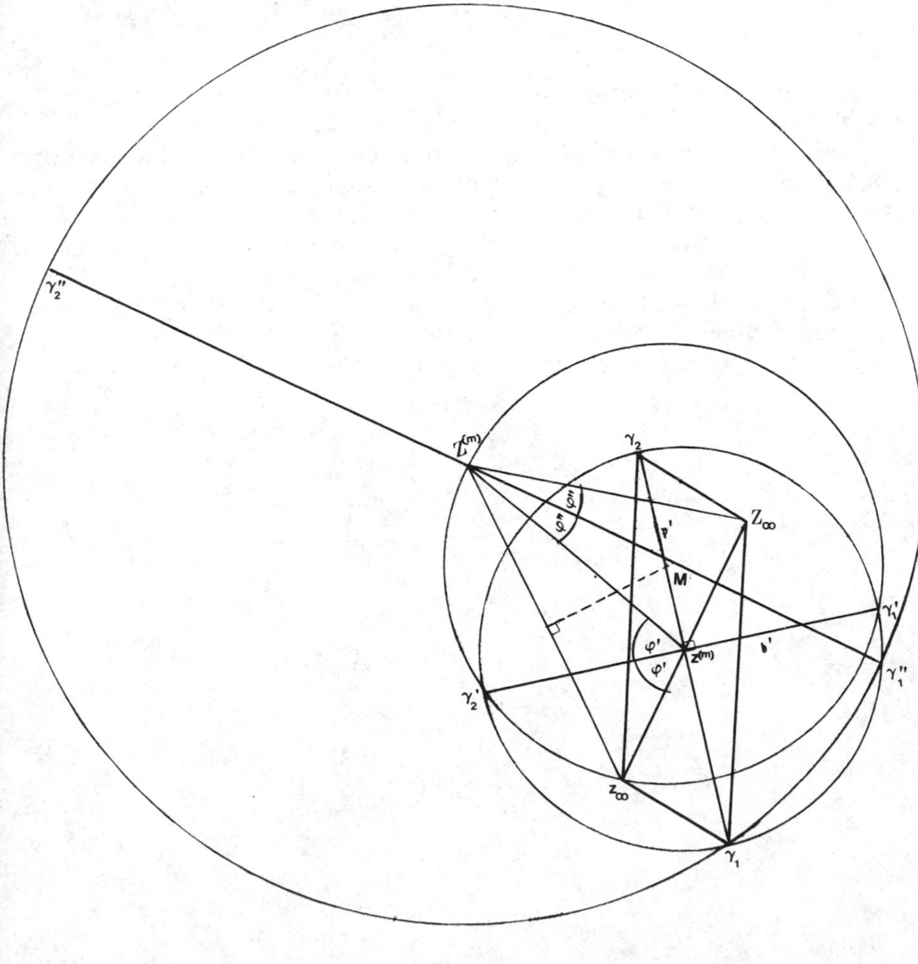

Fig. 34

$\mathfrak{b}'$ at $z^{(m)}$ and the perpendicular median to the segment $\mathfrak{s}(z_\infty, Z^{(m)})$. Both lines meet at a point M'. Draw the circle about M', passing through $Z^{(m)}$. It will meet $\mathfrak{b}'$ at the points γ_1' and γ_2'. The points γ_1 and γ_2 are the intersections of $\mathfrak{p}'$ with the circle about $z^{(m)}$ passing through γ_1' and γ_2'. This concludes the construction of the characteristic parallelogram of $\mathfrak{H}$.

The second factor $\mathfrak{Z}''$ of the decomposition of $\mathfrak{H}$ has $z^{(m)}$ and Z_∞ as a pair of conjugate points and the bisectrice of the angle $\sphericalangle z^{(m)}Z^{(m)}Z_\infty$ intersects the circle about $Z^{(m)}$, passing through γ_1 and γ_2, at the points γ_1'' and γ_2''. Then indeed

$$\mathfrak{Z}'(z_\infty) = Z^{(m)}, \qquad \mathfrak{Z}''(Z^{(m)}) = \infty, \qquad \mathfrak{Z}'(\gamma_1) = \gamma_2, \qquad \mathfrak{Z}''(\gamma_2) = \gamma_1$$

which defines $\mathfrak{H} = \mathfrak{Z}''\mathfrak{Z}'$.

Examples

1. Carry out the construction of Figure 31 in the case of a non-integral parabolic Moebius transformation $\mathfrak{H}$, for instance $\mathfrak{H}(z) = -z/(z - 1)$.
2. For a given characteristic parallelogram of $\mathfrak{H}$ find the parallelogram of $\mathfrak{T}\mathfrak{H}\mathfrak{T}^{-1}$ by geometric construction if the characteristic parallelogram of the non-integral Moebius transformation $\mathfrak{T}$ is prescribed.
3. Show that in Figure 33 the two triangles $\Delta\, \gamma_2 z_\infty'' z_\infty'$ and $\Delta\, z_\infty \gamma_1 z_\infty'$ are similar.
4. Generalize the construction of a decomposition of a non-integral Moebius transformation $\mathfrak{H}$ as indicated in Figure 33 by choosing the point z_∞' in an arbitrary position.

APPENDIX 4

COMPLEX NUMBERS IN GEOMETRY BY I. M. YAGLOM

The book by Yaglom appeared in Russian [1] in 1963, in an extended English translation [2] in 1968; it cannot be considered as an alternative for the present book as one might be led to believe by its title. A good deal of the geometry as well as of the algebra are different in both books. In our book we have restricted ourselves to the field $\boldsymbol{C}$ of the 'ordinary' complex numbers. Their identification with the points of a plane lends itself naturally to an analytic development of the geometry of the group of the circle preserving transformations (Moebius transformations and anti-homographies, cf. §§ 6–9) and some of its subgroups (§§ 13–15). Yaglom, however, introduces apart from $\boldsymbol{C}$ two other associative and commutative complex number systems as algebraic tools for the development of geometrical theories, namely

(a) The system $\boldsymbol{C}_0$ of the *dual numbers*

$$z = u + \varepsilon v, \qquad u, v \text{ real}, \quad \varepsilon^2 = 0, \quad \varepsilon \neq 0,$$

not a field, but a commutative ring with a unique unit element $1 + \varepsilon 0$ and with zero divisors $0 + \varepsilon v$. These dual numbers are identified with directed (oriented) straight lines in the plane, that is, u and v are taken as line coordinates. Again fractional-linear transformations

$$Z = \frac{az + b}{cz + d}, \qquad ad - bc \neq \varepsilon g; \quad a, b, c, d, g \text{ in } \boldsymbol{C}_0,$$

are analytic representations of geometrically meaningful mappings (e.g., euclidean motions of the plane onto itself) which carry lines onto lines and every four lines touching a circle or passing through a point onto four lines touching a circle or passing through a point, thus also mapping circles onto circles. Hence they are called 'axial circular transformations.' In another form they have been studied by E. Laguerre and are therefore called Laguerre transformations. As a four-line invariant under such transformations the cross ratio of four dual numbers has also a geometrical meaning. As in the case of $\boldsymbol{C}$ some striking geometrical theorems are obtained with the algebra $\boldsymbol{C}_0$.

(b) The system $\boldsymbol{C}_1$ of the so-called *double numbers*

$$z = u + ev, \qquad u, v \text{ real}, \quad e^2 = 1, \quad e \neq \pm 1,$$

a commutative ring with a unique unit element $1 + e0$ and divisors of zero

$u + eu$ (in fact $(1 + e)(1 - e) = 1 - e^2 = 0$). This system plays a role in hyperbolic (Lobačevski) geometry, similar to the role of C_0 in euclidean line geometry.

The work of W. Benz [2] is devoted to far-reaching generalizations of the basic ideas of Yaglom's book.

BIBLIOGRAPHY

J. ACZÉL (Ya. Ačel')

[1] Some general methods in the theory of functional equations in one variable. New applications of functional equations (in Russian). *Uspehi Mat. Nauk* (N.S.), **11**, no. 3 (69) (1956), 3–68. Cf. *Math. Reviews*, **18**, p. 807.

D. M. ADELMAN

[1] Note on the arithmetic of bilinear transformations. *Proc. Amer. Math. Soc.*, **1** (1950), 443–8.

F. BACHMANN

[1] Eine Kennzeichnung der Gruppe der gebrochen-linearen Transformationen. *Math. Annalen*, **126** (1953), 76–92.

[2] *Aufbau der Geometrie aus dem Spiegelungsbegriff.* (Sammlung "Grundlehren", Bd. 96), Berlin, 1959.

R. BALDUS

[1] Nichteuklidische Geometrie. Hyperbolische Geometrie der Ebene. *Sammlung Göschen*, **970** (Berlin-Leipzig, 1927).

H. BEHNKE and F. SOMMER

[1] *Theorie der analytischen Funktionen einer komplexen Veränderlichen* (Grundlehren, Bd. 77, Berlin, 1955) Kap. IV, §§ 3–5.

W. BENZ

[1] Über Möbiusebenen, Ein Bericht. *Jahresbericht der deutschen Mathematiker vereinigung*, **63** (1960), 1–27.

W. BLASCHKE

[1] *Vorlesungen ueber Differentialgeometrie III. Differentialgeometrie der Kreise und Kugeln* (Grundlehren, 29, Berlin, 1929).

R. BRAUER

[1] A characterization of null systems in projective space. *Bull. Amer. Math. Soc.*, **42** (1936), 247–54.

N. G. DE BRUIJN and G. SZEKERES

[1] On some exponential and polar representations of matrices. *Nieuw. Arch. Wisk.* (3), III (1955), 20–32.

C. CARATHEODORY

[1] *Conformal representation.* Cambridge Tracts, **28** (1932).

[2] The most general transformations of plane regions which transform circles into circles. *Bull. Amer. Math. Soc.*, **43** (1937), 573–9. (Gesammelte Mathematische Schriften, vol. III, **73** [München, 1955]).

[3] *Funktionentheorie, Band I* (Basel, 1950).

[4] *Theory of functions of a complex variable* (translation of [3] by F. Steinhardt) (New York, 1954).

E. CARTAN

[1] *La théorie des groupes finis et continus et l'analysis situs. Mémor. Sci. Math.*, **42** (Paris, 1930).

[2] *Leçons sur la géométrie projective complexe. Cahiers scientifiques*, **10** (Paris, 1931).

W. B. CARVER

[1] The conjugate coordinate system for plane euclidean geometry. *Amer. Math. Monthly*, **63** (1956). (Slaught Memorial Paper, no. 5, 86 pp., 1956).

A. Cayley

[1] On the correspondence of homographies and rotations. *Math. Annalen*, **15** (1879), 238–40. Coll. Math. Papers X, 153–4.

[2] On the matrix $\begin{pmatrix} a & b \\ c & d \end{pmatrix}$ and in connection therewith the function $\frac{ax+b}{cx+d}$. *Mess. of Math.* **9** (1880), 104–9. Coll. Math. Papers XI, 252–7.

S. Cohn-Vossen, *see* Hilbert and Cohn-Vossen

J. L. Coolidge

[1] *A treatise on the circle and the sphere* (Oxford, 1916).

[2] *The geometry of the complex domain* (Oxford, 1924).

C. Cosnita

[1] Sur une substitution homographique. *Bull. Inst. Politchn. Bucuresti*, **18** (1956), no. 3–4, 89–97. Cf. *Math. Reviews*, **20**, no. 3487 (1959), p. 577.

R. Courant

[1] *Differential and integral calculus* (London, 1937).

R. Courant and H. Robbins

[1] *What is mathematics?* (Oxford University Press, 1945).

H. S. M. Coxeter

[1] *Non-euclidean geometry*, Math. Expositions 2 (Toronto, 1957).

[2] *The real projective plane* (New York, 1949).

[3] On subgroups of the modular group, *J. Math. pures et appl.* (1958), 317–19.

[4] *Introduction to Geometry*, (New York, 1961), chapters 6, 9, 14, 16.

G. Darboux

[1] Sur le théorème fondamental de la géométrie projective. *Math. Annalen*, **17** (1880), 55–61.

R. Deaux

[1] *Introduction à la géométrie des nombres complexes* (Bruxelles, 1947).

[2] Sur l'image d'une affinité dans le plan de Gauss. *Mathésis*, **59** (1950), 101–10.

[3] A Moebius involution (advanced problem). *Amer. Math. Monthly*, **60** (1953), 127–8.

[4] Sur trois homographies du plan de Gauss. *Bull. de l'école polytechnique de Jassy*, **2** (1947), 106–16. (Cf. *Zentralblatt für Mathematik*, **32** [1950], 114.)

[5] Couples communs à une involution de Moebius et à une inversion isogonale. *Mathesis*, **63** (1954), 216–18.

[6] *Introduction to the geometry of complex numbers* (translation of [1] by H. Eves) (New York, 1957).

M. P. Drazin

[1] A note on permutable bilinear transformations. *Math. Gazette*, **36**, no. 315 (1952), 30–2.

P. Erdös and G. Piranian

[1] Sequences of linear fractional transformations. *Michigan Math. J.*, **6**, no. 3 (1959), 205–9.

H. Eves and V. E. Hoggatt

[1] Hyperbolic trigonometry derived from the Poincaré model. *Amer. Math. Monthly*, **58**, (1951), 469–74.

G. Ewald

[1] Axiomatischer Aufbau der Kreisgeometrie. *Math. Annalen*, **131** (1956), 354–71.

K. Fladt

[1] Bemerkungen zur Darstellung der ebenen hyperbolischen Geometrie im ebenen hyperbolischen Kreisbündel. *Acta. Math. Acad. Sci. Hung.*, **8** (1957), 99–105.

L. R. FORD

[1] *Automorphic functions* (New York, 1929).

H. G. FORDER

[1] Coordinates in geometry, *Auckland Univ. College Bull.*, **41**, Math. Ser. no. 1 (1953), 32 pp.

A. R. FORSYTH

[1] *Theory of functions* (Cambridge, 1900), p. 717.

R. FRICKE and F. KLEIN

[1] *Vorlesungen ueber die Theorie der automorphen Funktionen*, vol. I (2nd ed., Leipzig-Berlin, 1926).

E. GALOIS

[1] Démonstration d'un théorème sur les fractions continues périodiques. Oeuvres mathématiques (Paris, 1951), pp. 1–8.

K. GOLDBERG

[1] Unimodular matrices of order 2 that commute. *Washington Acad. Sci.*, **46** (1956), 337–8.

TH. GOT

[1] *Propriétés générales des groupes discontinus. Mémor. Sci. math.*, **60** (1933).

[2] *Domaines fondamentaux des groupes fuchsiens et automorphes. Mémor. Sci. math.*, **68** (1934).

G. H. HARDY

[1] *A course in pure mathematics* (Cambridge, 1938), chapter III.

D. HILBERT

[1] *Grundlagen der Geometrie* (6th ed., Leipzig-Berlin, 1923), Anhang IV (*Math. Annalen*, **56** [1902], 381–422).

D. HILBERT and S. COHN-VOSSEN

[1] *Anschauliche Geometrie* (Berlin, 1932).

[2] *Geometry and the imagination* (translation of [1]) (New York, 1952).

A. J. HOFFMAN

[1] A note on cross ratio. *Amer. Math. Monthly*, **58** (1951), 613–14.

J. E. HOFMANN

[1] Ueber sich nicht treffende hyperbolische Gerade. *Archiv der Math.*, **9** (1958), 219–27.

[2] Zur elementaren Dreiecksgeometrie in der komplexen Ebene, *Enseignement Math.* (2), **4** (1958), 178–211.

E. JACOBSTHAL

[1] Fibonaccische Polynome und Kreisteilungsgleichungen. *Sitzungsberichte der Berliner Math. Gesellschaft*, **17** (1919–20), 43–57 (cf. *Jahrbuch Fortschritte der Math.*, **47** [1924], 109).

[2] Zur Theorie der linearen Abbildungen. *Sitzungsberichte der Berliner Math. Gesellschaft*, **33** (1934), 15–34.

[3] Ueber die Klasseninvariante ähnlicher Abbildungen. *Kon. Norske Vid. Selskab Forhandlinger*, Part I, **25** (1952), 119–24; Part II, **26** (1953), 10–15.

[4] Ueber die Kreise, die durch eine gegebene lineare Funktion auf einen konzentrischen Kreis abgebildet werden. *Kon. Norske Vid. Selskabs Skrifter*, **3** (1954), 22 pp.

G. JULIA

[1] *Principes géométriques d'analyse* I. *Cahiers scientifiques*, **6** (Paris, 1930).

B. de Kerékjártó

[1] A geometrical theory of continuous groups I, II. *Ann. Math.*, **27** (1925-6), 105–17; *Ann Math.*, **29** (1928), 169–79.

[2] Sur le groupe des homographies et des anti-homographies d'une variable complexe. *Comm. math. Helv.*, **13** (1940), 68–82.

[3] Sur le caractère topologique du groupe homographique de la sphère. *Acta math.*, **74** (1941), 311–41 (*J. Math. pures et appl.*, **21** [1942], 67–100).

F. Klein

[1] Vergleichende Betrachtungen ueber geometrische Forschungen (Erlanger Programm 1872). *Math. Annalen*, **43** (1893) (Gesammelte Mathematische Werke Bd. I, no. 27, 460–97)

[2] *Vorlesungen ueber nicht-euklidische Geometrie* (Berlin, 1928).

[3] *Elementarmathematik vom höheren Standpunkt aus Bd. II: Geometrie* (Berlin, 1925).

[4] Zur Interpretation der komplexen Elemente in der Geometrie. *Göttinger Nachrichten* 1872, or *Math. Annalen* **22**, or *Gesammelte Mathematische* Werke Bd. I, no. 23, 402–5.

K. Kolden

[1] Continued fractions and linear substitutions. *Archiv for Math. og Naturvid. B.L.*, **6** (Oslo, 1949), 46 pp.

K. Leisenring

[1] A theorem on non-loxodromic Moebius transformations. *Michigan Math. J.*, **6** (1959), 51–52.

P. Libois

[1] Systèmes linéaires de projectivités entre deux formes de première espèce. *Mathésis*, **40** (1930), 121–8.

S. Lie and G. Scheffers

[1] *Geometrie der Berührungstransformationen* (Leipzig, 1896), pp. 414–6.

F. Löbell

[1] Eine Konstruktion des Punktepaares, das zu zwei gegebenen Punktepaaren der komplexen Zahlenebene harmonisch liegt. *Jahresber. Deutsche Math. Vereinigung*, **36** (1927), 364.

A. I. Markushievich (Markuševič)

[1] *Teoriya analitičeskich funkcii* (Moscow, 1950), chapter II, § 4.

V. Medek

[1] Linearne systemy projectivnych pribuznosti na priamke (Linear systems of projective transformations on a straight line) (Slovakian). *Mat. Fyz. Časopis Slovensk. Akad. Vied* 6, **2** (1956), 98–108. Cf. *Math. Reviews*, **18** (1957), 329.

R. Mehmke

[1] Zur Bestimmung des Punktepaares, das im Sinne von Moebius zwei gegebene Punktepaare der Ebene harmonisch trennt. *Jahresber. Deutsche Math. Vereinigung*, **37** (1928), 333–4.

A. F. MOEBIUS

[1] Ueber eine Methode, um von Relationen, welche der Longimetrie angehören, zu entsprechenden Sätzen der Planimetrie zu gelangen. *Leipziger Ber., math.-phys. Kl.*, **4** (1852), 41–54 (*J. f. Math.* [Crelle], **52** [1856], 229–42, and *Werke Bd. II*, 191–204).

[2] Ueber eine neue Verwandtschaft zwischen ebenen Figuren. *Leipziger Ber.*, **5** (1853), 14–24 (*J. f. Math.*, **52** [1856], 218–28, and *Werke Bd. II*, 205–17).

[3] Ueber die Involution von Punkten in einer Ebene. *Leipziger Ber.*, **5** (1853), 176–90 (*Werke Bd. II*, 219–36).

[4] Theorie der Kreisverwandtschaft in rein geometrischer Gestalt. *Leipziger Ber.*, **7** (1855), 529–95 (*Werke Bd. II*, 243–314).

P. MONTEL

[1] *Leçons sur les familles normales de fonctions analytiques et leurs applications* (Paris, 1927), chapter VIII.

F. MORLEY and F. V. MORLEY

[1] *Inversive geometry* (London, 1933).

F. MORLEY and J. R. MUSSELMAN

[1] On $2n$ points with a real cross ratio. *Amer. J. Math.*, **59** (1937), 787–92.

A. PANTAZI

[1] Sur certaines configurations d'homographies planes. *Bull. Math. Phys. École polytechn. Bucarest*, **7** (1937), 15–19.

D. PEDOE

[1] *Circles* (London, New York, Paris, 1957).

O. PERRON

[1] *Die Lehre von den Kettenbrüchen* (Leipzig, 1929).

E. PICARD

[1] *Leçons sur quelques équations fonctionelles avec des applications. Cahiers scientifiques*, **3** (Paris, 1928).

G. PIRANIAN, *see* P. ERDÖS

G. PIRANIAN and W. J. THRON

[1] Convergence properties of linear fractional transformations. *Michigan Math. J.*, **4** (1957), 129–35.

G. DE B. ROBINSON

[1] *Foundations of geometry*. Math. Expositions 1 (Toronto, 1946).

HERMANN SCHMIDT

[1] *Die Inversion und ihre Anwendungen* (München, 1950).

H. SCHWERDTFEGER

[1] Moebius transformations and continued fractions. *Bull. Amer. Math. Soc.*, **52** (1946), 307–10.

[2] *Introduction to linear algebra and the theory of matrices.* (Groningen, 1950).

[3] Zur Geometrie der Moebius-Transformation. *Math. Nachrichten*, **18** (1958), 168–72.

[4] On the discriminant $x'Ax \cdot y'Ay - (x'Ay)^2$. *Can. Math. Bull.*, **1** (1958), 175–9.

B. SEGRE

[1] Gli automorphismi del corpo complesso ed un problema di Corrado Segre. *Atti Accad. Naz. Lincei 1947 Rendiconti*, **3** (1947), 414–20.

C. SEGRE

[1] Note sur les homographies lineaires et leurs faisceaux. *J. r. a. Math.*, **100** (1887), 317–30.

F. SIMONART

[1] Sur les déplacements dans le plan complexe. *Acad. Roy. Belgique Bull. Cl. Sci.*, **38** (1952), 885–91.

C. STEPHANOS

[1] Mémoire sur la représentation des homographies binaires par des points de l'espace avec application à l'étude des rotations sphériques. *Math. Annalen*, **22** (1883), 299–367.

E. STUDY

[1] Das Appolonische Problem. *Math. Annalen*, **49** (1897), 497–542.

[2] Eine neue Art geometrischer Konstruktionen. *Sitzungsber. d. Niederrheinischen Gesellsch. f. Natur-u. Heilkunde* (1897), 1–7 (cf. E. A. Weiss [2]).

P. SZASZ

[1] Ueber die Trigonometrie des Poincaréschen Kreismodells der hyperbolischen Geometrie. *Acta Math. Acad. Sci. Hung.*, **5** (1954), 29–34.

[2] Hyperbolische Trigonometrie an dem Poincaréschen Kreismodell abgelesen. *Acta Math. Acad. Sci. Hung.*, **7** (1956), 65–69.

[3] Neuer Beweis für die Darstellung der Bewegungen und Umwendungen der hyperbolischen Ebene mit Hilfe der Hilbertschen Endenrechnung. *Ann. Univ. Sci. Budapest Eötvös Scet. Math.*, **1** (1958), 67–70.

O. TAUSSKY and J. TODD

[1] Commuting bilinear transformations and matrices. *J. Washington Acad. Sci.*, **46** (1956), 373–5.

W. J. THRON, *see* G. PIRANIAN.

J. TITS

[1] Généralisation des groups projectifs basée sur leurs propriétés de transitivité. *Mém. Acad. Roy. Belgique, Cl. Sci.*, **27** (1952), 1–115.

[2] Groupes triplement transitifs continus; généralisation d'un théorème de Kerékjártó. *Compos. math.*, **9** (1951), 85–96.

O. VEBLEN and J. W. YOUNG

[1] *Projective geometry I* (Boston, New York, 1910, 1938).

H. WAADELAND

[1] Ueber die Klassen ähnlicher linearer Abbildungen. *Kon. Norske Vid. Selskab Forhandlinger*, Part I, **25** (1952), 125–8; Part II, **25** (1952), 129–30.

G. N. WATSON

[1] A bilinear transformation. *Edinburgh Math. Notes*, no. 40 (1956), 1–7.

E. A. WEISS

[1] Zur Konstruktion des Punktepaares, das zu zwei gegebenen Punktepaaren der komplexen Zahlenebene harmonisch liegt. *Jahresber Deutsche Math. Vereinigung*, **37** (1928), 334–5.

[2] E. Study's Mathematische Schriften I. *Jahresber. Deutsche Math. Vereinigung*, **43** (1933), 108–24.

SUPPLEMENTARY BIBLIOGRAPHY

J. ACZÉL

[2] *Lectures on functional equations and their applications* (New York and London: Academic Press, 1966), pp. 232–3.

J. ACZÉL and M. A. MCKIERNAN

[1] On the characterization of plane projective and complex Moebius transformations. *Math. Nachrichten*, **33** (1967), 315–37.

R. ARTZY

[1] *Linear geometry* (Reading, Mass.: Addison-Wesley, 1965).

W. BENZ

[2] *Vorlesungen über Geometrie der Algebren* (Berlin, Heidelberg, New York: Springer-Verlag, 1973).

H. S. M. COXETER

[5] The inversive plane and hyperbolic space. *Abhandl. Math. Seminar*, Universität Hamburg, **29** (1966), 217–42.

[6] The Lorentz group and the group of homographies. *Proc. Internat. Congress on the Theory of Groups*, held at the Australian National University, Canberra, 1965 (1967), 73–7.

[7] Parallel lines. *Can. Math. Bull.* (1979).

H. S. M. COXETER and S. L. GREITZER

[1] *Geometry revisited*, New Math. Library 19 (New York: Random House, 1967), chapter 5.

P. J. DAVIS

[1] *The Schwarz function and its applications* (Carus Math., monograph no. 17, Math. Association of America, 1974).

H. EVES

[1] A survey of geometry (London: Allyn and Bacon, 1965).

J. GIBBONS AND C. WEBB

[1] Circle-preserving functions of spheres. *Trans. Amer. Math. Soc.* **248** (1979), 67–83.

H. HARUKI

[1] On the principle of circle transformation of a linear rational function in analytic function theory. *Duke Math. J.*, **36** (1969), 257–59.

[2] A characteristic property of orthogonal pencils of coaxial circles from the standpoint of conformal mapping. *Annales Polon. Math.*, **31** (1975), 171–7.

H. LIEBECK

[1] The convergence of sequences with linear fractional recurrence relation. *Amer. Math. Monthly*, **68** (1961), 353–5.

W. MAGNUS

[1] *Noneuclidean tesselations and their groups* (New York and London: Academic Press, 1974).

D. PEDOE

[2] *A course of geometry* (Cambridge University Press, 1970).

H. SCHWERDTFEGER

[5] Invariants of a class of transformation groups. *Aequationes Math.*, **14** (1976), 105–10.

[6] Invariants of a class of transformation groups II. *Aequationes Math.*, **17** (1978), 292–4.

R. STARROST

[1] *Das charakteristische Parallelogramm einer Moebiustransformation Staatsexamensarbeit* (Kiel, 1966), 53 pp.

I. M. YAGLOM (JAGLOM)

[1] *Complex numbers and their applications in geometry* (in Russian) (Gosudarst. izdat., Moscow, 1963).

[2] *Complex numbers in geometry* (translation of [1] by E. J. F. Primrose) (New York and London: Academic Press, 1968).

INDEX

A CATALOGUE OF SELECTED DOVER BOOKS
IN ALL FIELDS OF INTEREST

AMERICAN ANTIQUE FURNITURE, Edgar G. Miller, Jr. The basic coverage of all American furniture before 1840: chapters per item chronologically cover all types of furniture, with more than 2100 photos. Total of 1106pp. 7⅞ x 10¾. 21599-7, 21600-4 Pa., Two-vol. set $17.90

ILLUSTRATED GUIDE TO SHAKER FURNITURE, Robert Meader. Director, Shaker Museum, Old Chatham, presents up-to-date coverage of all furniture and appurtenances, with much on local styles not available elsewhere. 235 photos. 146pp. 9 x 12. 22819-3 Pa. $6.00

ORIENTAL RUGS, ANTIQUE AND MODERN, Walter A. Hawley. Persia, Turkey, Caucasus, Central Asia, China, other traditions. Best general survey of all aspects: styles and periods, manufacture, uses, symbols and their interpretation, and identification. 96 illustrations, 11 in color. 320pp. 6⅛ x 9¼. 22366-3 Pa. $6.95

CHINESE POTTERY AND PORCELAIN, R. L. Hobson. Detailed descriptions and analyses by former Keeper of the Department of Oriental Antiquities and Ethnography at the British Museum. Covers hundreds of pieces from primitive times to 1915. Still the standard text for most periods. 136 plates, 40 in full color. Total of 750pp. 5⅜ x 8½. 23253-0 Pa. $10.00

THE WARES OF THE MING DYNASTY, R. L. Hobson. Foremost scholar examines and illustrates many varieties of Ming (1368-1644). Famous blue and white, polychrome, lesser-known styles and shapes. 117 illustrations, 9 full color, of outstanding pieces. Total of 263pp. 6⅛ x 9¼. (Available in U.S. only) 23652-8 Pa. $6.00

Prices subject to change without notice.

Available at your book dealer or write for free catalogue to Dept. GI, Dover Publications, Inc., 180 Varick St., N.Y., N.Y. 10014. Dover publishes more than 175 books each year on science, elementary and advanced mathematics, biology, music, art, literary history, social sciences and other areas.